T0235463

Favorite Sons

Favorite Sons

The Politics and Poetics of the Sidney Family

Elizabeth Mazzola

First published 2003 by
PALGRAVE MACMILLAN™
175 Fifth Avenue, New York, N.Y. 10010 and
Houndmills, Basingstoke, Hampshire, England RG21 6XS.
Companies and representatives throughout the world.

PALGRAVE MACMILLAN is the global academic imprint of the
Palgrave Macmillan division of St. Martin's Press, LLC and of
Palgrave Macmillan Ltd. Macmillan® is a registered trademark in the
United States, United Kingdom and other countries. Palgrave
is a registered trademark in the European Union and other countries.

ISBN 978-1-349-73188-6   ISBN 978-1-137-09158-1 (eBook)

DOI 10.1007/978-1-137-09158-1

Library of Congress Cataloging-in-Publication Data

Mazzola, Elizabeth.
    Favorite sons: the politics and poetics of the Sidney family /
Elizabeth Mazzola.
        p. cm.
    Includes bibliographical references (p.) and index.
        1. Sidney, Philip, Sir, 1554–1586—Political and social views.
2. Politics and literature—Great Britain—History—16th century.
3. Pembroke, Mary Sidney Herbert, Countess of, 1561–1621—
Criticism and interpretation. 4. Wroth, Mary, Lady, ca. 1586–
ca. 1640—Criticism and interpretation. 5. Politics and literature—
Great Britain—History—16th century. 6. Politics and literature—
Great Britain—History—17th century. 7. Sidney, Robert,
1563–1626—Criticism and interpretation. 8. Poets, English—Early
modern, 1500–1700—Biography. 9. Great Britain—Politics and
government—1558–1603. 10. Sidney, Philip, Sir, 1554–1586.
Arcadia. 11. Sidney, Philip, Sir, 1554–1586—Family.
12. Sidney family. I. Title.

PR2343.M29 2004
821'.3—dc21   2003048820

A catalogue record for this book is available from the British Library.

Design by Newgen Imaging Systems (P) Ltd., Chennai, India.

First edition: October, 2003
10 9 8 7 6 5 4 3 2 1

For Jim, Pamela, and Alison

CONTENTS

ACKNOWLEDGMENTS

I must first thank my students at The City College of New York, who so patiently listened to every incarnation of my ideas about families, the Sidneys, and literary traditions outlined here. They have always proved a good-natured and generous audience, and the book is better because it took its first steps under their watch. I am also indebted to my colleagues at City College, especially Paul Oppenheimer, who has been a constant source of inspiration, advice, and interest throughout the writing of this book. My chairman J. Fred Reynolds obtained for me a teaching release that allowed me to finish this book; I am extremely grateful both to him and to Dean James Watts for their continued support of my research and scholarly life. I wish I could better thank the late Charles T. Mark for his friendship. I also acknowledge the generous help provided by the Rifkind Foundation of The City College and by The City University of New York PSC-CUNY Research Award Program, which provided me with a summer research grant. Just as crucial to my writing has been the assistance of the Interlibrary Loan Department of the Cohen Library at City College, particularly the help of Richard Uttich and Evelyn Bodden, who were relentless in their efforts to track down research materials. I am grateful, too, to the reading room staffs of the Henry E. Huntington Library and Pierpont Morgan Library. Finally, I want to acknowledge the continuous support of my editor Amanda L. Johnson and her assistant Matthew Ashford.

The extraordinary work of Margaret Hannay, Maureen Quilligan, Gary Waller, and the late Josephine A. Roberts has been invaluable to this study and my debts to these scholars extensive. There are other scholars and friends scattered near and far whose encouragement and thinking were a mainstay in other important ways: I especially thank Ernest B. Gilman, Julie Eckerly, Peter Herman, Corinne S. Abate, Selma Erhardt, and Robert E. Stillman for their support of this project. Above all, I am grateful to Clare Kinney and Mary Ellen Lamb. Their readings of Sidney and Wroth have been essential to this project, and their assistance immediate, generous, and invaluable: they shared their work, read my own with alacrity and enthusiasm, and their golden touch refined many clumsy steps. Simply put, they taught me how to think about this book: whatever seems right benefited from their counsel, and any mistakes are mine entirely.

Parts of this study were presented at a 2001 Special Session of the Sidney Society at the International Congress of Medieval Studies at Kalamazoo. Other parts have had other audiences. I thank Erika Gaffney, Ann Donahue, and Ashgate Publishing Ltd. for allowing me to reprint part of chapter 1, which appears as " 'Natural' Boys and 'Hard' Stepmothers" in Corinne S. Abate's essay collection, *Privacy, Domesticity, and Women in Early Modern England* (2003); Renata Wasserman and Wayne State University Press, for permission to reprint portions of chapter 2, which originally appeared as "Brothers' Keepers and Philip's Siblings" in *Criticism* 41, 4 (1999): 513–42; and Joseph Black, who allowed me to reprint part of the epilogue, which originally appeared as "Spenser, Sidney, and Second Thoughts" in the *Sidney Journal* 18, 1 (2000): 57–81.

There are other debts Milton describes—immense ones of endless gratitude—but these are debts I can happily always owe, for in the course of writing this book about families I have counted the blessings of my own, and to my husband and children this book is dedicated, with love.

INTRODUCTION
"A NOBILITY NEVER INTERRUPTED"

The vicissitudes of family life really belong to history, just as the peculiar charms and snares of families ultimately comprise culture at large. Perhaps that is the most telling—and most easily overlooked—feature of family life, that its fortunes and failures cannot help but transpire in a larger world, one more defective or limited and finally more forbidding than the seemingly closed world of parents and children, brothers and sisters, husbands and wives. Family history is driven by and experienced through the logic of culture.[1] But families, I argue here, are poetic projects too: they have an internal logic, an ongoing if unspoken scheme for inviting affection and posting reward, for orienting members, drawing them into its circle while obscuring the world outside. The family straddles culture and biology, politics and eros.[2] But it can only pretend to evade such awesome forces. Indeed, the family is neither a refuge nor a playground but a conflicted region crowded with all of the conditions of representation, the chief setting for mischief and passion, the first site and subject of all thought.[3]

The tension and pliancy and durability of family connections are features with which family members continually experiment, sometimes in the midst of more pressing activities, sometimes at a loss for other opportunities to act.[4] But the members of Sir Philip Sidney's family, as I explore here, repeatedly turned to poetry as a vehicle to analyze or to extend familial connections, sometimes intending to map out their patchwork pattern of values and concerns, at other times roughly to plough through them. Their efforts are partly explained by a Renaissance sensibility that understands poetry as both a private and eminently public expression, as a vehicle for prominently displaying inner thought. But another part is due to the transformed shape of the family in the early modern period. As historians like Lawrence Stone have variously chronicled, the idea of the family prevalent through the Middle Ages, defined and governed by extensive ties of kinship, was starting to contract and center itself, with the result that a once-wide network of cousins, godparents, ghosts and in-laws slowly gave way to a narrower circle of immediate family members, parents and their children.[5] Anthropologist Jack Goody similarly explains that the gradual disappearance of extensive patrilineal descent groups rendered the bilateral core of kinship—a circle of fathers and sons—the dominant form.[6] Such radical revisions were marked by new legal arrangements, physical spaces, sexual practices, and emotional habits, by the new ways that a psyche might construe or rebel against authority, and by the new stories that might be told about families.[7]

These developments—or restrictions—of family life in the early modern period only seem clear-cut. The family's spheres of interest are psychological and economic, influencing one's relation to the world as well as what one makes of it.[8] If the family is a material structure it is a metaphorical one as well,[9] with subtle complications and ambiguities at work in the laws of kinship presiding over its small circles of dependency and obligation and interest. For one thing, as the family became a more isolated enterprise, it also found itself increasingly subject to official sanctions, protections, and surveillance, transformed into a private region cramped by all of the apparatus of state and Church. The family's new private and public status is illustrated in Mary Sidney Herbert's 1601 letter to Elizabeth Tudor, where the countess of Pembroke construes her family's pliancy and durability as useful instruments of state. The countess describes her relation to the queen in terms of a "more particuler bond," and then proceeds to dedicate her twenty-one-year-old son William, "partisepating of the same sprite," to the queen's service. "I do as gladly leve him and give him as ever I was made mother of him[,]" she explains: "And acordingly am to take Comfort in him as he shall be blessed in yowr gratious sight and frame him self wholly to please and serve yowr most Exelent Majesty."[10]

Familial allegiances go hand in hand with transfers of power, amended calculations of interest, and a redistribution of affection; indeed, the countess's mothering instincts illustrate the struggles and consolations intrinsic to Renaissance family life, making clear that this is a system open to a reworking of loyalty and shifting of influence, even with a mother's blessing. Natalie Zemon Davis characterizes early modern family life in "terms of strategy, identity, and order," but I would emphasize too its elastic quality: perhaps it was less a "privileged locus for identity, gratification, and reward" than a site of ready transformation or household magic as "[l]awful," Shakespeare's Leontes exclaims, "as eating" (*The Winter's Tale* 5.3.111).[11] Actually, Shakespeare takes up this theme again and again in his work, exploring how the

Renaissance family might continually redraw the lines between insiders and outsiders and collapse the thin walls separating the household from the world outside.[12]

* * *

Such familial revisions were often public though not always so theatrical. The members of the Sidney family proudly represented themselves in terms of their famed lineage even as they construed their lineage as something eminently accommodating or flexible, even fictive, merely now "illuminated" by what Stone calls the "dying embers of lineage loyalty."[13] I argue that their model for such a practical—if artful—consciousness was supplied by the legendary Sir Philip Sidney (1554–86), celebrated throughout Europe as a warrior, statesman, and Protestant hero. This heroic image was only part of Sidney's legacy to his family, however. Sidney also created a set of fictions about himself as reluctant poet, compelled into writing and into retirement by the intransigence of the larger world, especially of his queen.[14] If Sidney's pose was a real one it was also an artifice designed to make something happen: Sidney's fiction about himself was instrumental and experiential at the same time, a mode of being and a medium through which he might alter his fate and be pressed into more fitting or satisfying service. Other fictions provided him with the chance to imagine for himself and his family "a nobility never interrupted."[15]

The catastrophe of Sidney's sudden death at the age of thirty-one transformed his suggestive image: around it now collected ideas about a poet's sincerity or ingenuity, ideas that would animate literary careers and literary history in England, ideas we still draw upon today. The disappointments of public life, for instance, came to emblematize the perfections of Sidney's private life, and Sidney's uncertainty about his "unelected vocation" as poet perfectly illustrated an individual sensibility at war with the world and at odds with himself. Sidney's anxious poetics ultimately render for us a picture of the poet. The ironies of his pose only multiplied in death, for then Sidney became public property, both prince of poets and pattern of poetry. Utterly accommodating and pliable, Sidney's image is both of absent father and favorite son, a formidable model whose mistakes could easily be forgotten and misgivings overlooked.[16] It is this mixed legacy, I argue, which spurred the efforts, literary and otherwise, of family members who survived him.

Other scholars have pointed to the opportunities Sidney's example supplied as well as to the anxieties his model induced. But no one has really explained what happened to Philip Sidney's legacy in the process of recovering it, or what shape the English literary tradition would assume with Sidney's troubled image as its impetus and original riddle. Yet his descendants' literary experiments with language, genre, and poetic structure were nearly always also struggles with Sidney's awkward example, efforts to get his ghost to stay put and to make sense.[17] Indeed, Lady Mary Wroth's *Urania* offers a way to read a world split in two by Sidney's "absent presence." As I argue in chapter 4, the romance of the *Urania* also tracks the movements of an ambivalent, beloved hero and supplies a cosmology through which to read the new world that takes shape around his loss.

* * *

The history of the nuclear family more or less finds its beginnings in the early modern period, and part of what we can recover in studying Sidney and Wroth is how individual family members gradually assumed larger proportions in the development of an individual psyche. Freud's speculations about the origins and motives of this psyche, for example, are only made possible through the Renaissance configuration of the family:

> The Oedipus complex, which begins with the prohibition of the child's sexual activity, falls within the emotionally charged context of parent–child relations in the privatized, nuclear family. Isolated in the nuclear family, the child's entire emotional life is centred on its parents, on their affection and hostility, on their autonomous power to set the rules of the child, on the depth of the identifications the child makes with them . . . The super-ego . . . assumes that the child will *forever be a member of the family*.[18]

To be sure, all families are the stuff, as Freud maintained, of dreams: cavernous repositories of secrets and fantasies, sites of painful deferrals and instant gratifications, founts of formidable lies and easy consolations. But there are important differences between Renaissance families and our own. Although a new model of family life was slowly emerging, the Renaissance family still frequently comprised a world of "universal nephews," brothers and kinsmen, sometimes "spurious children" and countless "cousins."[19] The sheer variety of metaphor testifies both to the elasticity and the attenuation of extended family ties in the early modern period,[20] and consequently to models of authority, privilege, obligation, rivalry, and expectation which can seem strange or cold or sometimes cruel. But families tie people together in manifold ways. Early modern family life was no less profound or demanding in terms of the ways it encourages emotion and material interest alike,[21] perhaps because its burdens were shouldered by many more people in an intricate and not always self-evident web of relation.

And for this reason, perhaps, the very first task imposed on Sidney's literary descendants was to figure out how to construe his authority, given his ambivalence about his labors. To take Sidney as one's model meant disregarding his deepest (or loudest) ideas about poetry and resuscitating something Sidney wanted buried with him. If my argument builds on earlier accounts of Sidney's complicated influence and of the Renaissance literary tradition as proposed by Richard Helgerson, Raphael Falco, Richard McCoy, and John Guillory,[22] it diverges with their researches in crucial ways, because I emphasize just how formative and vexing Sidney's model was. I also seek to read Sidney's heirs like Mary Wroth, Mary Herbert, and Robert Sidney as self-conscious mythographers keenly aware of the problems Sidney's work posed for their own. In this way, I locate in the Renaissance family a powerful scaffolding for history, myth, and cultural authority, arguing that the afterlife of Philip Sidney's brief, luminous career emblematizes the way early modern families could preserve their values by reinventing themselves.

* * *

This book traces the dimensions, fictions, and contradictions of Sidney's life as poet, and also considers how family members in Sidney's wake sought to appropriate and correct his flawed example. As Lawrence Lipking notes, "no poet becomes himself without inheriting an idea of what it means to be a poet."[23] This also suggests that a poetic inheritance is always in flux, never a fixed set of riches or resources, something also true of families and their wealth. My study is at once psychological and historical in its treatment of the early modern family as a complex and ongoing system of power and privilege, affection and rivalry, a vehicle for presumption and desire, at once a center and a margin where tradition collided with experiment. As Stone proposes, the "Open Lineage" model characterized by a larger kin network and lukewarm relations was gradually replaced more or less by 1550 with a nuclear family presided over by a powerful patriarch.[24] With a decline of loyalties to kin, patrons, and local community, Stone argues, there also arose more "universalistic loyalties" to nation and Church, institutions which might make harsher or more sweeping demands upon their members.[25]

In this new world, family secrets might remain hidden even as they provided a "consistent internal logic" shaping psychic history and cultural productions, including poetry and poetic careers.[26] Family secrets could even be concealed by other stories, other myths. Certainly the early modern family is as much an engineering miracle of detachment as it is a mechanism for affection.

My ideas about early modern families are clearly indebted to the work of social historians like Stone and Peter Laslett, as well as to the findings of psychoanalysts like Freud and Melanie Klein and D.W. Winnicott. The historians I draw on suggest how early modern culture molded the family in its image, while the psychoanalysts I consult explore how inner worlds provide the ground rules for outer ones—so that siblings become almost as important as parents to one's development in the early modern period. Another consequence of linking history and psychology is the picture of families as entities bound by culture. This means that families take their shapes in history and in imaginations, and thus that no matter how rigid or unexplored their rules or sentiments are, families continually flourish, adjusting and finding their satisfactions within the world at large.

The relation between families and the worlds they inhabit is then a mutual one, and my book analyzes how inner fictions contribute to outer ones: how myths shape families and how families in turn manifest culture in its most authoritative form, generating (or, more often, simply maintaining) ideas about gender, morality, love, and other fictions. I argue that aristocratic families in particular might weather insults on their structure through the stories they told about themselves, narratives of virtue (or class) rewarded, histories of promotion and punishment, poems about apparent failure. Sidney illustrates the close (or closed) reading of families in the Old *Arcadia* (1580), where the doddering patriarch Basilius hopes to rewrite his family's history by adhering to the advice of an oracle. The early modern family was not merely a microcosm of culture but, as Sidney suggests here and elsewhere, culture's earliest and most intense setting; if Basilius's Arcadian retreat offers a refuge from some pressures, it serves as a "hot-house" for others.[27] No wonder Sidney explains his poetry as shaped by its confines, the product of "his not old years and idlest times."

*   *   *

Scholars like Mary Ellen Lamb, Gary F. Waller, and Josephine Roberts have ably described the accomplishments of the members of the so-called Sidney "circle," a literary network with Philip as its head (or figurehead) and his sister Mary Sidney Herbert (1561–1621), patron, writer, executor of Sidney's literary estate, and guardian of his image. In addition, Philip and Mary's brother Robert (1563–1626) produced a manuscript of poems (only recovered in the 1970s); and Robert's daughter Lady Mary Wroth (1587?–1653?) authored *The Countess of Montgomery's Urania* (1621)—the first work of prose fiction written in England by a woman—along with a sonnet sequence and a play.[28] Members of Sidney's family took control of Sidney's image even as they interpreted it in diverse ways, reworking lyric poetry, Protestant poetics, and the romance in his wake. The fledgling English literary tradition, I argue, therefore offers a "hot-house" atmosphere similar to the one provided by the early modern family,[29] crowded with patrons and offshoots and by near and far-flung members of the Sidney circle, including figures like Edmund Spenser, Ben Jonson, and Wroth's cousin William Herbert.

I focus here on a handful of the most deliberate and self-conscious members of this circle whose writings seek to elaborate and perfect Sidney's problematic example, especially Wroth, who shares her uncle's anxiety about the early modern family. The romance fictions of uncle and niece are concerned primarily with what Lamb has nicely called the "genealogical dilemma" posed by the Sidney family.[30] Indeed, Wroth advances her claims to literary authority and familial legitimacy by revising the work of her uncle, taking seriously poetic projects he had regarded with tremendous irony. Yet both figures devote enormous narrative energies to pictures of families in distress, organized by more complex politics and sublimations and by fewer rewards as they become increasingly isolated from the larger world. I argue, too, that both Sidney and Wroth understand their vocation ("unelected" or not) in moral terms, in that it met a requirement to make things cohere: to continue—sometimes by transforming—their family's understanding of itself.

I am not solely interested in what Gary Waller has so fruitfully called the "Sidney family romance," however.[31] By focusing on the dimensions of Sidney's legacy as well as on the structure of the early modern family, I also want to explore what might be called the Sidneian genealogy of romance, the genre of desire organized by and through the ceaseless riddles of family life. The romance worlds outlined by Sidney and Wroth illuminate a claustrophobic universe (however densely populated) of sameness and familiarity, where morality and narrative are closely aligned because both exemplify family codes and paternalistic structures. Peter Gay has reminded us that "the psychoanalyst's individual is the social individual,"[32] and his identification of familial ambitions with erotic drives on the one hand and with social mechanisms on the other—his insistence, in other words, that the historian and the psychoanalyst share a common interest—is valuable to my efforts to consider the early modern family as a primary social locus and literary effect. At once, the early modern family supplies the grammar of reality and a vocabulary of desire. For both Wroth and Sidney, it is also a source of early anxieties and lasting satisfactions—a setting for shared privacy and joint interiority. And for both writers, this narrow and crowded region is the world of romance.

* * *

This book is divided into two sections. The first part, "Domestic Arrangements and Household Privileges," analyzes Sidney's family's place in the world of the court and how the court, in turn, sought to make itself a home to affection and prestige, desire and reward. I also analyze how aristocratic families like the Sidneys construed themselves in a universe where lineage and affect, deference and authority were almost hopelessly tangled. Chapter 1 investigates Sidney's conflicted relationship with the queen and explores how their difficult association characterizes one between early modern mothers and children more generally. The queen employed a maternal rhetoric shot through with mixed messages of power and tenderness, and many of her courtiers responded with the juvenilia of tournaments, youthful pastoral songs, and the cautious innovations of love poetry. Drawing on psychological accounts of child development supplied by Freud, Klein, and Winnicott, I examine how Sidney seeks to rewrite the rules of this maternal dynamic by reinventing infantile power and constructing his own language of the family, aiming to win the queen's love without remaining her child.

Chapter 2 continues my study of Renaissance fictions of the nursery by taking up the work of Sidney's siblings. If his brother and sister both begin to write after Sidney's death, they address the problem of familial identity in entirely different ways. Mary Herbert completes Sidney's psalm translations, exploiting the psalmist's rhetoric of anonymity so as to fashion a language of Protestant memory that might permit her own poetic exertions. Robert Sidney plumbs his brother's erotic poetry in contrast and supplies a radically different picture of the life of the poet: alienated, not detached; isolated physically, not mentally. Despite these stark differences in how they understand affective claims and poetic callings, Mary and Robert's strong misreadings of their brother's legacy prove more formative, I argue, than even Philip Sidney's poetry. The first part of the book treats the codes and values of emerging family life as shaping early modern poetry. The second part, "Compromising Positions and Rival Romances," analyzes how early modern prose, in turn, arranges and interprets the sentiments, debts, and secrets organizing the emerging family. If the first part of the book also focuses on what C.L. Barber calls "vertical relationships," ones that "originate in the transmission of heritage and identity,"[33] the second part treats what might be called "horizontal relationships," familial ties that reinforce relation or assume likeness by privileging connection over authority. This new kinship structure is encouraged and expressed by early modern prose, a far more effective exponent and arbiter of the new ways in which culture arranges and shares it values—not through reifying them but by displacing or extending them.

Chapter 3 reads Sidney's Old *Arcadia* as an elaborate investigation of the privileges and defects of early modern families. Typically, familial ties are thinned or confused or temporarily overturned through romance designs capitalizing on the genre's anxieties about rape and incest. But Sidney turns to prose in his romance as a vehicle more suited than poetry to preserve the fragile claims of Renaissance families and rescue them from erotic confusion, reviving a nearly outmoded genre in the process. Theorists Wlad Godzich and Jeffrey Kittay have proposed that the medium of prose materializes out of familial disruptions and affective dislocations like the ones Sidney catalogs;[34] and Sidney uses prose, I maintain, to mitigate the threat of incest and smooth over the disruptions of exogamy by weaving new patterns of filiation.[35]

Chapter 4 focuses on Lady Mary Wroth as the most ambitious reader of Sidney's legacy. Recent critics seem particularly concerned to rescue Wroth from Sidney's shadow,[36] but I argue that Wroth's twinned interests in genealogy and romance repeatedly draw on her uncle's writings, and that her prose in the *Urania* seeks to relieve familial anxieties as Sidney's prose does in the *Arcadia*.[37] Yet Wroth makes more explicit than her uncle the ways romance's values and narrative devices are closely aligned, both derived from exhausted family codes and frayed paternal structures. Chapter 4 also builds on the work of theorists of the novel like Michael McKeon and Nancy Armstrong who describe the early novel in terms of its transformed heroics, physical settings, and forms of relation. Wroth belongs in such accounts of the novel, I suggest: the *Urania* reconfigures families and their claims by taking up the stories of slighted younger brothers, typically abject figures who—along with pirates, thieves, and other prodigals—provide the early novel with a new grammar of reality, vocabulary of desire, and economy of sentiment.

In the epilogue I briefly consider Spenser's (and later poets') treatment of the Sidney myth. Spenser analyzes Sidney's fractured image repeatedly in *The Faerie Queene* as well as in shorter poems, especially *Muiopotmos* (1591) and *Astrophel* (1595), and I would propose that his entire career was marked by sustained meditation on Sidney as both "president" of chivalry and failed poet. Spenser's sentiments seem particularly ambivalent in *Muiopotmos*, where Edenic pleasures are recovered as infantile ones, heroic virtue is displayed as immaturity, and Sidneian genealogy is revealed to be a dead end.

* * *

In exploring how storytelling advances the aims of early modern families and culture at large, my book also aims to complicate the reach and meaning of cultural authority. For one thing, what can be known in family circles and in the recesses of the imagination may overlap yet be irreconcilable and discontinuous. There is no complete embrace in these regions fashioned by affection and power, no matter how sorely craved or carefully cultivated. That there are arenas of thought and feeling where separate people can nonetheless love and work and think together is something Sidney and Wroth movingly propose, even if their romance fictions are finally skeptical of the endurance or satisfactions of such familial unions. To be sure, Wroth is more deliberate in the way she questions the limits and premises of connection, her prose a vehicle for refusing history much in the way Pamphilia's sonnets to Amphilanthus constantly refuse his love. In later decades, novelists like Aphra Behn, Daniel Defoe, and Samuel Richardson will actively seek mechanisms for generating other forms of relatedness and new metaphors of family. Wroth's stories of sisters and brothers, favorite sons and slighted daughters are prescient in their reminders that childhood is not the only place where the family is so formative—so necessary to, and generative of, stories of belonging and escape.

Part I

Domestic Arrangements and Household Privileges

CHAPTER 1

"NATURAL" BOYS AND "HARD"
STEPMOTHERS

The situation of non-satisfaction in which the amounts of stimulation
rise to an unpleasurable height . . . must for the infant be analogous to
the experience of being born.[1]

Your faire mother is a bed,
Candles out, and curtaines spread:
She thinkes you do letters write:
Write, but first let me endite:

(Fourth Song, *Astrophil and Stella*)[2]

Infancy has a variety of narratives, few of them tender. Breast-feeding
becomes an occasion, for instance, for destructive impulses, anxiety,
sadistic biting: even the seemingly contented baby may be impaired by
an "excess of oral satisfaction."[3] If such accounts of childhood are always
shaped by adult rationalizations (so that, we learn, "although but little
aggression may be observable it is not possible to ignore the destructive
element in the aim of the infant"[4]), then these mature stories demonstrate how flawed all development is, how much growth is founded on
error. We are always in a state of being born. This may also suggest that
we are always at war with our mothers, an insight the childless Elizabeth
Tudor frequently used to her political advantage.[5]

Her nation might someday have a wiser ruler, the queen once told
her people, but no one would love them better. "And though you have
had and may have many princes more mighty and wise sitting in this
seat," she stated in a 1601 address to Parliament, "yet you never had nor
shall have any that will be more careful and loving."[6] In a culture of
manifold fictions and symbols, this picture of exclusive "loving" was the
queen's preeminent "shaping fantasy," her symbol of herself.[7] But the
poetry that exploited Elizabeth's symbolism had at once to sustain and
numb this symbol's power, for the queen's kind of mother's love had its
own singular rewards and punishments—the magical realm it fashioned

also built out of calculated indifference, willing subjugation, and often obscure victory. In Elizabeth's nursery, childish pleasure was political, and therefore inconstant, unreliable, and short-lived.

Because acknowledgment trumps knowledge in the mother's kingdom, affection is a narrow secret, so vulnerable because so unlike other forms of reason. D.W. Winnicott's picture of the world shared between mother and child takes this epistemological difference into account, proposing the same maternal qualifications and abilities Elizabeth had outlined for her people:

> A *mother* need not have intellectual understanding of her job because she is fitted for it in its essentials by her biological orientation to her own baby. It is the fact of her devotion to her own baby rather than her self-conscious knowledge that makes her good enough to be successful in the early stages of infant nurture.[8]

However, as Winnicott later notes, while the mother's "biological orientation" equips her for responsibilities toward her child, the child experiences her physical presence as something *psychological,* in terms of the powerful feelings which frustrate or content him.[9] Winnicott's—or Elizabeth's—shrewd distinction between wisdom and care would therefore mean little to those whom the queen promises to love. Indeed, such opacity is the burden of Elizabeth's subjects, made partners in a loving game that presumed their inability to know better.

Philip Sidney attempts to rewrite the rules of this game, as I hope to show in this chapter: he seeks to win the queen's maternal love without remaining her child. One example of this difficult maneuvering occurs explicitly in the apology that opens his sonnet sequence *Astrophil and Stella.* "LOVING in truth, and faine in verse [his] love to show, / That the deare She might take some pleasure of [his] paine," the self-conscious sonneteer Astrophil is unable to submit to conventional strategies for winning his beloved by making her his nurse or his teacher, so that her "Knowledge might pitie winne, and pitie grace obtaine" (1.1–2, 4). Astrophil cannot occupy the standard role of lover reduced to frustrated baby for, "great with child to speake" (12), he suffers the pangs of labor too. Seeking "fit words" and "[s]tudying inventions fine," the pain Astrophil experiences is literary anxiety, not the disease of infancy, because he seeks to be interpreted, not to be silenced. And if his anxiety appears regressive—"Biting [his] trewand pen, beating [him] selfe for spite" (13)—Astrophil nevertheless is able to substitute his efforts for those of a mother. Indeed, he assumes Elizabeth's professed limitations and plays at playing dumb, simply looking in his heart and writing.

Although Astrophil regularly positions himself against poetic rivals as figures of insincerity or immaturity ("Enam'ling with pied flowers their thoughts of gold" [3.4], their "rimes, running in ratling rowes" [15.6]), he also suggests the puerile if perilous quality of the love he craves from his beloved. The scene in Astrophil's study recapitulates many of the terrors of the nursery, where the mother can seem a monster and even the reach of the child's body can become a source of distress. But in the Fourth Song we learn that Astrophil and his beloved can "in their best language woo" when the mother is held "Dumbe" with "sleepe" (16, 26). Under these relaxed conditions Astrophil is free to acknowledge Stella's talents too, telling her: "Write, but first let me endite" (40).[10]

We might use the details of Sidney's biography as a guide to Astrophil's complicated representations of intimacy in the nursery, for the "faire mother" of Penelope Rich (the model for Stella) was Lettice Knollys, Sidney's uncle Leicester's second wife. When Sidney is presumed to be writing the sonnets, sometime in 1581–82, Lettice had just given birth to the son who displaced Sidney as Leicester's heir. The same figure who thus encourages Astrophil's talents in the shape of Stella produced a rival who dashed all of Sidney's dynastic hopes.[11] I would propose that Stella's mother and Stella herself are two aspects of Sidney's queen, one a nurturing force who threatens the poet with disfavor, the other a generally unresponsive reader with her own powers to hate, or to write.

* * *

While sheer generational conflict is the subject of many Renaissance narratives,[12] a more awkward discrepancy between adult aspiration and childish want (or between instruction and delight) is almost as often taken up in early modern texts. More's *Utopia* and Machiavelli's *Prince* provide two examples of the disjunction between urbane invention and rude necessity. The split is pushed far apart in Shakespeare's *Lear*, when Cordelia cannot force her heart into her mouth during the king's speaking contest and flatter her father. Indeed, there are numerous instances in *Lear* where parents and children are at odds or where mature reasoning deflects immature desire. A shamed Gloucester, for example, is unable to reconcile his worldly knowledge with a lawless appetite—his fatherhood, in fact, something only supported by his learning. We see the educational process still occurring when he introduces his bastard son Edmund to Kent: "I have so often blushed to acknowledge him," Gloucester discloses, "that now I am brazed to't" (1.1.8–10).

At first, Kent doesn't grasp Gloucester's rueful admission about his son or about the hardened condition of Gloucester's body and he tells the father, "Sir I cannot conceive you." But Gloucester reassures him: "Sir, this young fellow's mother could." This formula for paternity really works the other way around: Edmund's nameless mother comprehends the father because she comprehended their son, her knowing another part of her "biological orientation," inconsequential because inevitable. Like the mother "a bed" in Astrophil's Fourth Song, Edmund's mother is a figure of authority exactly because she is dumb.

The "facts" of "devotion," as Winnicott terms them, shape other facts, larger epistemologies. This is because the first schoolroom, the nursery, has its own array of techniques for construing knowledge, whereby affection can silence reason or ignore sons "by order of law" like Edmund's brother Edgar. We would be wrong to overlook these early sites of ill-disciplined learning or uncoordinated thought. We would be wrong because even "bad" ideas have an important place in any system or culture. The philosopher Gregory Bateson supplies them with the same weight as good ideas in his account of the "double bind" a child can find himself trapped in, when the right answer is just as inappropriate or threatening as a wrong one can be. Caught in this "no-win" situation "in which the other person . . . is expressing two orders of message and one of these denies the other," "[t]he child is punished for discriminating accurately what [the mother] is expressing, and he is punished for discriminating inaccurately—he is caught in a double bind."[13]

Bateson illustrates his premise with the scenario of a child whose overwhelming need for love disarms the mother and whose seeming refusal of it angers her. Another example is provided by Astrophil's continued retractions in the sonnets. Sidney's spokesman indicates the problem of the double bind most clearly when he spells out how epistemological and affective purposes are in conflict in his verse to Stella, his reasoning always in "weake proportion" to his love: "So that I cannot chuse but write my mind, / And cannot chuse but put out what I write" (50.7, 9–10). If his awareness of the problem is acute, Astrophil is still unable to resolve it, and instead finds himself backsliding "[w]hile those poore babes their death in birth do find" (11), again and again forced to relearn how to speak and feel.

Every occasion for intimacy always risks the possibility of incoherence, but such occasions were multiple at Elizabeth's court—Sidney's pastoral entertainment *The Lady of May* a famous instance where the queen obtusely failed to carry out the part the poet had carefully scripted for her. "Privilege" and "privation" could be rudely yoked in early modern England,[14] where status supplied proximity to the queen even as wealth underwrote sites far removed from public scrutiny or approval.[15] The connections between privilege and privation flourished, too, in Elizabeth's princely assurance of a mother's tenderness, promising to love her subjects because they did not question her ability to do so.

Of course, all royal subjects must yield to their rulers: the difference was that Elizabeth pretended to yield to her people as well. As a result, her courtiers found themselves in the "double binds" Bateson describes throughout her long reign, baffled by the queen's mixed messages of power and tenderness, virginity and desirability, remoteness as an icon and frailty as a woman. In their dismay and confusion, many of Elizabeth's subjects advanced their careers through the juvenilia of tournaments and "cautious innovations" of love poetry, in youthful pastoral songs and apologies, and the doubtful affirmations of allegory.[16] If the queen instructed her "favorites" to remain infants, she made infantile desire acceptable, even artful, behavior.

* * *

Before I explore Sidney's reinvention of infantile power—even as he chose (or was required) to employ the queen's symbols and affects—I want to examine the sometimes crude dynamics of Elizabeth's court at greater length. Bateson's picture of the schizophrenic child's environment may seem an unusual analogy to the political and psychological exigencies of the Tudor world; yet his model of how loving mothers mandate dissimulation and enforce humiliation writes large the vulnerable and volatile state of the nursery, a setting where privilege and privation are perhaps nowhere else more closely bound.

Bateson claims that the schizophrenic child is not physically imprisoned by his parents but held captive by their logic, the child's knowing completely governed by adult desires when the mother controls "the child's definitions of his own messages, as well as the definition of his responses to her."[17] Astrophil outlines such imaginative constraints when he points to the limits of his skill along with those of his audience, complaining: "WHAT may words say, or what may words not say, / Where truth it selfe must speake like flatterie?" (35.1–2). No wonder every effort to communicate in Bateson's universe becomes a paralyzing gesture, every attempt at reason transformed into a dangerous error. This is a world that brooks no agreement, much like the brilliant cosmos over which Elizabeth ruled for forty-five years. The consequences for ignoring such a mistress could be just as severe as serving her, and Spenser, Ralegh, Essex, and Sidney's careers all reflect the foreshortened perspective of the nursery, where ambition had to be curtailed since devotion proved always inadequate.

To be sure, the conflated picture Elizabeth presented of herself as queen and mother, virgin and lover inspired a *range* of feeling, combining slavish admiration and witty innuendo with trivial self-aggrandizement. If a child's relation to his parents evolves then his conflicted feelings of hate and desire and awe will sort themselves out. But works of art do not banish immaturity or insist on clear-cut articulations of emotion: they may instead cultivate psychological limitation; and in Elizabethan poetry we have a miscellany of reactions to maternal power and stratagems for reimagining its tensile force which never completely dislodge it, or really even aim to. When Bateson details the pathology that "[occurs] in the human organism when certain formal patterns of the breaching occur in the communication between mother and child,"[18] we can make out the patchwork image of Elizabeth, veiled by nervous myth making and enveloped by crippled courtiers.[19] The queen not only insisted on the stunted exertions and "ornate capitulations"[20] Bateson describes but made them a cultural policy, the preferred way to conduct business or to produce art.

*   *   *

Under Elizabeth, courtly poetry and politics alike were transformed into vehicles for wooing a capricious, despotic beloved, someone who would survive aggression and withstand rage and remain beautiful, desirable, and entirely in possession of her courtiers' faculties.[21] This pathology is diagnosed by Joseph Loewenstein as a *peri bathous* or "art of sinking," molded by a culture of steady repression, boredom, and repetition.[22] One of its most famous artifacts was presented by a chastened Sidney when he made a 1581 New Year's gift to his queen of a pin coated with jewels and shaped like a whip. Sidney provides Elizabeth with an elegant means to punish him and elevate herself, along with an obviously public statement about the inevitability of this transaction.[23] The witticism cannot override Sidney's humiliation, it only adorns it.

Such debilitating arts could be required of other queens, too, as when Mary Stuart was invited to yield her maternal role so that Elizabeth might better carry out her own. While imprisoned in England, the Scottish queen was asked to persuade her son James VI to cooperate with Elizabeth's advisors, and Elizabeth's Secretary of State Francis Walsingham was likewise instructed, "You shall then [in your meeting with Mary] as of your self lay before her the inconveniency, and danger that may grow both to her self and her son in case she shall show herself any way unwilling to employ her credit and authority with her son for the performance of those said offers."[24] If the best mothers ultimately deny themselves for their children's advancement, then Elizabeth and Bateson insinuate how the best children will mortify themselves for their mother's pleasure.

Keith Thomas and Joseph Loewenstein have explored the unusual demographics of an early modern world grown dangerously juvenile. As Thomas explains, "the prevailing ideal was gerontocratic: the young were to serve and the old were to rule. Justification for so obvious a truth was found in the law of nature, the fifth commandment, and the proverbial wisdom of the ages."[25] But the new wisdom redefined what it meant to be young, too. The thirty-four-year-old Robert Sidney was told, for instance, that his queen "thought him *too* young for any place about her"[26] (my emphasis).

Drawing on the work of Loewenstein and Thomas might help us to explore the aesthetics of this world, and help us to see how Sidney's writings struggle to think through the "double bind" by persuading his queen to acknowledge his adult position. It would seem that Sidney was only partly successful. For one thing, Petrarchan poetry more typically functioned as a set of mnemonic devices or "commemorative fictions" which signaled a poet's faulty vision: it was hardly the means to clarify what a lover could say to his beloved or what a baby might teach his nurse.[27] At the same time, the queen was increasingly becoming an "antique image"; once "the subject of painters" she was now more of "an object on which paint would be applied."[28] The rewards of Elizabethan art and politics became increasingly premised, in other words, on the artist's inability to see or tell. His "young braine captiv'd in golden cage" (23.11), poets like Astrophil are both a threat to and balm for this system, flatterers who urge their readers not to take them seriously. Their instruction can require disabling effort. If Sidney invents Astrophil, Astrophil will in turn invent a pathetic double, telling Stella, who "cannot skill to pitie [his] disgrace," "I am not I, pitie the tale of me" (45.3, 14).

* * *

While reason, ambition, and the New World were compressed by the parameters of the queen's gaze—made into mirrors in which she might "behold" or remake her face (*The Faerie Queene*, proem to Book 2)— more and more courtiers were confined to the world of Elizabeth's nursery, where they could remain safely ignored or overruled.[29] A sizeable portion of Elizabethan poetry was written by subjects who remained "young" and on the margins of power, including figures like Sidney and Spenser (1552–99), and players even further removed from court, like Robert Southwell (1561–95), Thomas Nashe (1567–1601), and Christopher Marlowe (1564–93).

Katherine Duncan-Jones comments on "the obvious point that all Sidney's poetry is *early* poetry"[30] but we might press her point harder. Certainly Petrarchism served many poets as an adolescent game, its pedantry competitive, its stifled sentiments not only passive but pacifying. Yet there were other immature sentiments, and Theresa Krier outlines alternative responses to Elizabeth's royal repressions. Reactions toward the queen's maternal figure ranged beyond mere gynephobia or misogyny, for in arrogating to herself so much of a mother's cultural authority, Elizabeth also inspired much of the anxiety which that image induces, inflaming fears about infantilism and dependence, animating longing and frustration. As Krier suggests, what emerged culture-wide was a dense "choreography" of gestures and movements toward this figure, "complex instances of identification and disidentification, acknowledgement and disavowal." Spenser's *Faerie Queene* utilizes many of these gestures and movements, combining allegory and romance to describe political and psychological frustration as well as to obscure it. Spenser represents the sovereign mother by splintering her into multiple and contradictory images, some glorious—like the virgin princess Una or warrior bride Britomart, and some menacing—like the emasculating Queen Radigund, or warrior bride Britomart.

Clearly the splitting of the maternal image was not always effective, even in Spenser's capable hands. "It is not accidental," Krier writes,

> that so much work on subject formation in Renaissance culture uses psychoanalytic theories of infancy and their complex negotiations with feminist theory. . . . Attention to the infant's subjective experience of the threshold between preoedipal and oedipal has become a powerful instrument suited to English culture and writing in the sixteenth century. . . . [given] the boy's sharp rupture from the feminine that was so often central to English Renaissance child-rearing, educational institutions, and professions.[31]

I would simply add that the boy's uneasy relation to his mother signaled additional tensions within English culture. The choreography Krier describes was staged not only in nurseries (where wet nurses and mothers took turns identifying and rejecting children) but in churches as well (where embattled Catholics and Protestants rescued their God's body from each other's symbols) and in the early modern theater (where audiences were educated about the rudiments of experience and learned which passions to identify with or expel).[32] "[F]amisht" and in an "Orphane place," Astrophil will conclude his sequence by characterizing Stella as an "ABSENT presence" (106.3,6,1), highlighting at once his imaginative autonomy and his despair at attaining it.

The harshness of such labor as well as the delight in the "choreography" Krier describes are things infants quickly discern in their energized immobility and eloquent blankness. Their knowing and feeling transparent to each other, their most profound relatedness is sparked by the idea that the mother has no inner life but belongs entirely to them. What they learn early on is that the one-sidedness of such intimacy can work both ways.

*   *   *

I now turn to consider the contentment and anxiety Sidney experienced as the queen's "foster child" or stepson.[33] Astrophil is the most concrete vehicle Sidney uses to formulate his undeveloped experience, a figure who insists on the "facts" of "devotion" with the determination of a child refusing more learned distinctions, "Desire still cr[ying], 'give me some food'" (71.14). Many critics take Astrophil's outbursts at face value and emphasize Sidney's shrill insistence of his needs rather than his eloquent analysis of them.[34] These readers vacillate only between treating Sidney's case as one of house arrest or of arrested development, unsure of whether Elizabeth was wise to curb his vast political ambitions or needlessly fearful about ambitions that were doomed to self-destruct.[35] Sidney's work deepens such confusion. This is because, like the gift of a jeweled whip, it employs deliberately emptied signs[36] to show how mothers and children humor each other.

Indeed, the jeweled whip is a neat metaphor for the cunning and strength of the double bind, a signal that Sidney knows of and knows how to succumb to infantile defeat. As I explore more fully in the following, Astrophil's sonnets similarly draw on Petrarch's fawning treatment of the beloved while transforming this talent into facile self-promotion.[37] Maureen Quilligan reminds us too, "in the midst of a prevailing fashion for courtly compliment . . . [Sidney's invention of Stella] may have been its most pointed aggression against [Elizabeth's] central authority. He would not play politics by her rules but would turn her Petrarchan forms to his own purposes."[38] Still, *Astrophil and Stella* is just one battle in a larger psychological war, and Elizabeth is only one of his opponents.

Sidney's poetics repeatedly traverse the ground between childish tantrum and polemic directed against himself, something we can also see in the two defenses he produced in his short career, like bookends demarcating public triumphs and private pains. Another example of such divided accomplishment is located behind the polite criticism Sidney inserts into a 1579 letter protesting the queen's rumored marriage to the Catholic Alencon. Sidney explains the queen's poor choice of suitor by reminding her of a royal defect, maintaining that, unlike the courtiers who monitor her whereabouts constantly, she cannot see herself. "Our eyes," he suggests, "[are] delighted in the sight of you . . . your own eyes cannot see yourself."[39] Sidney hints that the delights of the nursery are not always reciprocal, that in fact Elizabeth might also be diminished by her unique positioning.

Shakespeare's Volumnia seems to comment on a similar dynamic when she tells her warrior son Coriolanus, "Thy valiantness was mine, thou sucked'st it from me" (3.3.129). Yet Sidney's "valiantness" never completely belonged to him, either: his renowned act of generosity on the battlefield, when the wounded hero gave his water bottle to a dying soldier, is likely a fiction. And there are other borrowings. The self-styled "pro-rex" of Ireland, Sidney was also Leicester and Warwick's heir and "a universal nephew figure" to a number of older humanists who lacked heirs.[40] As his childhood friend Fulke Greville sums up, Sidney was "a great lord by birth" but also by "alliance and grace."[41]

Greville is especially sensitive to the rougher edges of Sidney's position. Describing his friend's reasons for drafting the risky letter to the queen, Greville explains,

> for Sir Philip, being neither magistrate nor counsellor, to oppose himself against his sovereign's pleasure in things indifferent, I must answer that his worth, truth, favour and sincerity of heart—together with his real manner of proceeding in it—were his privileges, because this gentleman's course in this great business was not by murmur among equals or inferiors to detract from princes, or by a mutinous kind of bemoaning error to stir up ill affections in their minds, whose best thoughts could do him no good, but by a due address of his humble reasons to the Queen herself. (37)

Greville details the unmatched assurance the humble Sidney expected to find in his queen through a privileged relation that would make every other tie with equals or inferiors—with anyone else, that is—unnecessary. But if Greville's account outlines an impressive bond then it also implies its enervating effect; and after the failure of the letter Greville seeks to reassure readers about Sidney's unimpeded abilities, inadvertently exposing his friend's weakened condition. Greville comments that Sidney now "seemed to stand alone, yet he stood upright" (38).

*   *   *

Likely undertaken after what Greville calls the "dangerous error" of the letter and Sidney's subsequent retirement from court (a period Sidney himself describes as his "not old years" and "idlest times"), *The Defence of Poetry* (1579–80) outlines a way for Sidney to be a poet without having to publish poetry, supplying a means to espouse his "unelected vocation" and at the same time escape from it, allowing him to own up to mature self-awareness and still remain a child.[42] Ostensibly his argument is that poetry leads to virtuous action and "take[s] naughtiness away,"[43] but poetry also makes allowances for childish behavior. The poet, Sidney claims, "is the food for the tenderest stomachs,"[44] and "men (most of which are childish in the best things)" "will be glad to hear the tales of Hercules, Achilles, Cyrus, Aeneas."[45] He describes poetry's gentle sustenance and epistemological value in contrast to the queen's unapologetic rules of engagement, calling it "the first light-giver to ignorance, and first nurse, whose milk by little and little enabled them to feed afterwards of tougher knowledges."[46] Moreover, only poetry promotes true affection between parents and children, for Sidney likens the "ungratefulness" of poetry's detractors to "vipers, that with their birth kill their parents."[47] Nevertheless, if being a child is how Sidney feels, it also serves as a way to express adult frustration, now that England, once "the mother of excellent minds" has become so "hard a stepmother to poets."[48]

Astrophil and Stella (1581–82) elaborates this picture of the poet as child by continually rewriting sexual desire in terms of infantile contentment.[49] Indeed, while Arthur Marotti reads Astrophil's erotic longings as a screen for Sidney's political ambitions (so that the queen was being courted in the sonnets solely for political favors), childish confinement appears something desired by the poet, at odds with, maybe suspicious of any adult ambitions.[50] As Astrophil explains it:

> LOVE still a boy, and oft a wanton is,
> School'd onely by his mother's tender eye:
> What wonder then if he his lesson misse,
> When for so soft a rod deare play he trie?
> And yet my Starre, because a sugred kisse
> In sport I suckt, while she asleepe did lie,
> Doth lowre, nay, chide; nay, threat for only this:
> Sweet, it was saucie *Love*, not humble I. (73.1–8)

Astrophil revises the scenario of the Fourth Song, since here the boy disturbs the sleeping mother to steal from her a "sugred kisse."

If the lovers are conflated with the conventional picture of Venus and her son cupid, Astrophil pushes this imagery still harder, using it to avoid responsibility and undercut ambition: "Sweet, it was saucie *Love*," he tells her, "not humble I." His regressive impulses are matched by ones that magnify Stella, and Astrophil warns others:

> But no scuse serves, she makes her wrath appeare
> In Beautie's throne, see now who dares come neare. (9–10)

But this reading of intimacy (with its playful rhetoric of insubordination: "see now who dares come neare") construes babies as helpless bullies and maternal wrath as seductive, for the emboldened child Astrophil concludes "[t]hat Anger' selfe I needs must kisse againe" (14). Later he avers: "I will but kisse, I never more will bite" (82.14). Astrophil's picture of infantile power can in fact be simultaneously generous and menacing: Stella is also a baby who in the Fift Song has been given "Ambrosian pap" (26), yet she is admonished that "Sweet babes must babies have, but shrewd gyrles must be beat'n" (36).

* * *

Alongside these complex consolations Sidney occasionally turned to the comfort of the tirade, or what Astrophil calls a "wailing eloquence" (38.11). Perhaps this relief is depicted in Greville's account of the tennis court insult. Apparently days after the letter to the queen,[51] the earl of Oxford (one of the party in favor of Elizabeth's nuptials) tried to force Sidney from the courts and Sidney refused, complaining that the "ill-disciplined" earl's lack of courtesy prevented his acquiescing.[52] Courtly combat quickly degenerated into something like a sandbox dispute, where an enraged Oxford called Sidney a puppy and Sidney retorted with an almost unintelligible genealogical riposte, telling Oxford: "all the world knows puppies are gotten by dogs and children by men" (39).[53]

A similar revisionary mythology is presented, however, in *The Defence of the Earl of Leicester* (1584–85), perhaps Sidney's last work and the only one intended for publication, written in response to the anonymous *Leicester's Commonwealth*, which appeared in the summer of 1584.[54] *Leicester's Commonwealth* saddled Leicester with all kinds of charges (many already familiar to its readers) including Leicester's involvement in his first wife's death and his ambition to marry Elizabeth and/or remove the queen from power. Ignoring these scandalous charges, Sidney instead strives to glorify the noble links he shares with his uncle, his aim to countenance the libel that Leicester "want[ed] gentry."[55]

Katherine Duncan-Jones claims that Sidney was "too anxious" in the *Defence of Leicester* "to show himself Leicester's 'loyal and natural boy'" after the death of his uncle's son in 1584.[56] Yet Sidney's earlier aim to deliver a "golden" world from Nature's "brazen" one as outlined in the *Defence of Poetry*[57] also fits into these larger ambitions to posit lineage and ratify affection. Sidney had elsewhere implied that the queen could be replaced by more satisfying forms of consolation, as he wrote to his father-in-law Walsingham, describing the economic perils afflicting Leicester's troops in the Netherlands:

> If her majesty were the fountain I would fear considering what I daily find that we should wax dry, but she is but a means whom God useth and I know not whether I am deceived but I am faithfully persuaded that if she should withdraw herself other springs would rise to help this action.[58]

This hypothesis of alternative sources of sustenance similarly motivates the strange designs and energies of *The Defence of Leicester*.[59] Just as much of Sidney's poetry tackles the tangled issues of the poet's autonomy and audience, his prose works wrestle with these problems as well (something I explore more fully in chapter 3). Only here Sidney's more private anxieties—a child's need for comfort, a courtier's need for promotion—were more publicly exhibited, justified, and allayed.

But Sidney's harried response to Leicester's libeller was never published. This is probably a good thing, since the work contains none of Sidney's masterful wit or irony, only sarcasm and self-regard and the petulance of someone who knows he's being looked at, buttressed, he claims, by the assurance of a "nobility never interrupted."[60] Sidney now upholds a patchwork model of a patriarchal order, a model which makes clear that even if all of Leicester's "great honours came to him by his mother,"[61] Leicester is clearly not dependent on the queen. Indeed, Elizabeth takes her glory from him: "[Leicester's] faith is so linked to her Majesty's service, that who goes about to undermine the one, resolves withal to overthrow the other."[62] In addition, Sidney proposes a revised vision of maternal power utterly divorced from the queen's royal authority by relocating it to ancient history. He reports that "even from the Roman time to modern times in such case they might, if they listed, and so often did, use their mother's name; and that Augustus Caesar had both name and empire of Caesar only by his mother's right."[63] In this way Sidney's propaganda for mothers becomes propaganda for himself.

* * *

If a mother first rescues the infant from extinction, her caring may foist a thicker oblivion on her baby, the way Titania imperiously tends the orphaned child of her votaress in Shakespeare's *Midsummer Night's Dream*. Titania's love and care can never symbolize or be replaced by another's, as the fairy queen assures Oberon, "Set your heart at rest / The fairyland buys not the child of me" (2.1.122). What better reminder in Titania's exclusive loving of what Stephen Greenblatt calls the "tragic inescapability of continuous selfhood"?[64] And what better monument to the strained compass of poetry, which strives again and again to compensate children finally forced from their mothers' arms?

We have another monument in the complicated picture of Sidney, the same intricate narrative of tenderness and oblivion, violence and indifference at work in accounts of his unsatisfying career and untimely death. Once his Grand Tour was concluded in 1575, the young Sidney served Elizabeth more or less as a "carpet knight": denied Henry Sidney's post in Ireland or any sustained diplomatic service abroad; succeeding his father only as ceremonial "Cup-bearer" to the queen in 1576; forced into "retirement" at his sister's estate in 1579/80; knighted solely as a courtesy to Prince Casimir in 1583; and prevented from accompanying Francis Drake on a 1585 expedition to the West Indies.[65] Interestingly, a good part of the works which comprise Sidney's magnificent oeuvre were composed during the brief period (1581–84) when he was not Leicester's heir and therefore most in need of the queen's favor.[66] After Sidney was finally permitted to accompany his uncle to battle against pro-Spanish forces in the Low Countries in November 1585, he died a few months later, wounded in a skirmish at the age of thirty-one. On his deathbed he begged that his work be destroyed.

Even Sidney's elegists contradict themselves, puzzled by the meaning of his short life and pointless death. In some accounts Sidney is a Protestant martyr who heroically threw off his cuisses to meet his unarmed opponent, in other accounts he had simply been too hasty to do battle and neglected to properly arm himself.[67] Moreover, Sidney's considerable frustrations as a courtier and the little pride he took in his "unelected vocation" as poet hold no place in contemporary stories about him.[68] What counts is the mangled body and the generous spirit, his defeat a poignant signal that the queen understood her subjects better than they did.[69]

In the aftermath of the tennis court incident, Greville recorded the queen's gentle rebuke of her subject after Sidney had rashly challenged Oxford to a duel. Greville's account experiments with the positioning of queen and subject, mother and child, alternating similes with other images of relation, contrasting figures of autonomy with those of dependence: "[L]ike an excellent monarch," the queen "la[id] before [Sidney] the difference in degree between earls and gentlemen; the respect inferiors ought to their superiors; and the necessity in princes to maintain their own creations" (41). Greville's picture couples princely affection with noble disinterest, maternal devotion with queenly equanimity.

The same scrupulous Elizabeth must have delighted in the tribute paid to her as queen and mother in *A Midsummer Night's Dream*, attended on by frail mortals who give her their children. To be sure, mothering is frequently rewritten by Shakespeare, brought into line with angry offspring, as when Hamlet tutors Gertrude, or magically redeemed by a kiss that renders the mother mortal, the way Hermione is stirred from art to life in *The Winter's Tale*. Spenser makes motherhood the ultimate wish fulfillment for poets and their nations alike, his figure of Britomart both lost child and great mother of England. But Sidney's biographer Greville supplies us with a drastically different image of maternal authority in describing Elizabeth's interview with Sidney: "although [Sidney] found a sweet stream of sovereign humours in that well-tempered Lady to run against him, yet found he safety in her self, against the selfness which appeared to threaten him in her" (37).[70] Perhaps Elizabeth's princely heart was less a "sanctuary" unto Sidney (Greville 38) than an opulent grave, her love the kind that reduced the men around her to children, though it might exalt some of them with poetry.

CHAPTER 2

BROTHERS' KEEPERS AND PHILIP'S SIBLINGS

I have to fill a double space. I have to be my brother as well as myself.[1]
Now I, now I my self forgotten find,
Even like a dead man, dreamed out of mind,
       (Philip Sidney, Psalm 31 "In Te, Domine, Speravi")

Even in the closest circles, Philip Sidney remained something of a riddle. Long before he died he was the subject of mythmaking, serving at once as "perfecte paterne of a Poet," emblem of courtly aspiration and Protestant hero, a beloved and important figure who in reality appears sadly ineffectual and frustrated. These frustrations were the subject of chapter 1; how and why Sidney's complicated example as poet flourished nonetheless is the subject of this one.

Sidney seems to have inspired the same kind of devoted misunderstanding in family members that he inspired in English culture at large. Yet this misunderstanding characterizes many Renaissance families where affective ties were often bound to abstract entities; and if Sidney was a shining example for English poets to emulate, he was also, Katherine Duncan-Jones maintains, a "universal nephew figure" for a number of older humanists who lacked heirs.[2] Something that served a variety of hopes and range of desires, Sidney's detachment was charismatic because it could so readily, comfortably, assuredly be employed. Sidney was curiously a step or two removed, too, from both the courtly and literary centers of activity that sought to take him as an exemplar. When he died in 1586 he was famous neither for his knightly success nor for his poetry; in fact, it is his self-proclaimed heirs, poets like Mary Herbert and Robert Sidney, Edmund Spenser or Ben Jonson for example, to whom we are indebted for our sense of Sidney's literary importance. Indeed, Jonson and Spenser advance their careers by praising him.

Deriving "influence" always involves backward glances, but Sidney's image seems less stable the further back we look; what flashes before us is a fugitive phantasm, something always contemplated from a distance.[3] If we locate his contributions by considering what later poets took from him, we observe that Sidney's legacy derives not from the bounty of work published after his death (since it was circulated during his lifetime) but from an ancillary or indirect influence, spread out along a mazy river, not part of any main stream. He serves as a kind of corollary that ran along rather than clearly organized English verse and provided a deeper, hidden structure that could underpin—or more subtly cloak—a poet's vocation. I believe this secret, powerful current is at its most secret and most powerful in the work of Philip's brother and sister, both of whom began to write after their brother's death. Their poetry keeps alive and safely hidden a fallen knight, a silent hero, a buried puzzle.

* * *

To be sure, unlike Spenser's self-effacing narrator (challenging us to beg his question "who knowes not *Colin Clout*?") or Shakespeare's chameleon-like persona, Philip Sidney's poems and prose boldly announce his voice in multiple ways, as ingenious lover Astrophil, as stern critic and divine singer, as his uncle Leicester's spokesman. Perhaps Sidney's literary impact was energizing because it was so flexible or pliant (even allowing later poets, as Raphael Falco suggests, to "invent" him as their precursor). Yet in some ways Robert and Mary were better positioned to make use of Philip's learning and example than he was because they were better hidden than him: Robert as *second* son, loving husband, and dutiful public servant, Mary as wife, mother, and executor of Philip's estate. I want to consider the way their poetry challenges and exploits their brother's originality and obscurity.

More broadly, I explore how culture employs some families to keep its secrets and how families manipulate culture to tell stories about themselves. Familial bonds in early modern England were fragile and loose and relatively weak, but they could be made to harness other forces, inviting outside threats or dissolving them, as we see in Elizabeth's charged relations with her courtiers. Another example of such techniques of "familiarity" was provided when Mary Tudor famously proclaimed Elizabeth a bastard after Mary assumed the throne in 1553. Of course, the queenly Catholic Mary's blood positioned her to judge her Protestant sister's lowly status, but historians like Lawrence Stone have noted the cool, relaxed ties between family members in other early modern households. Because of mortality rates and practices like primogeniture and fosterage, family members might be closer to neighbors or friends than to each other, aligning themselves with kin in terms of fear or need rather than through love. Their emotional detachment may have enabled the family to weather medical or economic storms, however, alleviating grief over an infant's death or deploying a workforce of adolescent apprentices (the "internalization" of family ties that encourages deeper, more intimate affections and desires may partly explain the problems tearing at the fabric of modern nuclear families).[4] The family might prove to be a more useful political or religious concept than a psychic tool. Still, there are gestures like Mary Tudor's where family bonds are important not because they have been internalized or encourage feeling but because they have become elaborately external or symbolic, established by complicated cultural codes or devised to provide solutions to larger problems.

We have an example of the way familial ties could be turned inside out and made unfamiliar in the way Mary exposes her illicit relation to her sister. Another example is found in one of Sir Henry Sidney's letters to his son, dated 1578. Here we encounter a rather odd piece of fatherly advice, for humanist insight is couched by a mix of admiration and recrimination. The elder Sidney (the Lord Deputy of Ireland) puts his son forward while pushing himself aside, and paternal instruction reworks some biblical commonplaces and commandments. But this is a letter to Henry's *second* son Robert, Philip Sidney's younger brother, and Henry is recommending Philip to Robert as a model. Perhaps because he is not addressing his first son and heir (and the heir as well of the earls of Warwick and Leicester), Henry had a bit more room to play with the "tropic consanguinities" Lena Cowen Orlin describes,[5] making fluid what is normally fixed by blood, telling a son to ignore Castiglione—as well as himself—in favor of his brother's example:

> [W]hat do I blunder at thyes thyngys, follo the dyrectyon of your most lovyng brother, who in lovyng you, is comparable with me, or exceldyth me. Imitate hys vertues, exercyses, Studyes, & accyons; he ys a rare ornament of thys age, the very formular, that all well dysposed young Gentlymen of ouer Court, do form allsoe thear maners & lyfe by. In troth I speake yt wythout flatery of hym, or of my self, he hathe the most rare vertues that ever I found in any man . . . Ons agayn I say Imytate hym.[6]

Although Henry's description of a son whose loving "exceldyth" his own provides Robert with an image to copy (indeed, "the very formular"), the unnamed Philip appears less an additional source of comfort than a rival to the "blundering" elder Sidney. Father to all "Gentlymen," this child's example condemns others to hollow rhetoric; as Henry concludes: "Ons agayn I say Imytate hym."

It is not only Henry Sidney whom Philip appears to silence. Very little has been written about Robert Sidney, husband, father, soldier, patron, and little-known Renaissance poet.[7] If much more information has been assembled about Robert's sister Mary Sidney Herbert, the countess of Pembroke, most of what we know about the English Renaissance reveals something about their brother Philip. But I want to demonstrate in the first part of this chapter that the fortunes and dimensions of the Sidney family can also tell us something about the Renaissance in England. In the second part, I turn to the fortunes and dimensions of Philip Sidney's shadowy image, so useful because it was so unwieldy.

*   *   *

Although there is no gene (at the moment) that spells out literary talent or verse-making ability, there exist significant literary families like the Brontes, the Jameses, and the Manns who seem to draw on a pool of literary resources or habits.[8] Still, in terms of political connectedness, religiosity, and the intangibles of nobility, the Sidneys are unusually well connected, their "family culture" remarkably alive even today.[9] What accounts, then, for Robert's middling literary success, given Mary's now-celebrated status as well as Robert's invaluable ties and the huge legitimating apparatus of titles, preferment, aristocratic pieties, and humanist accomplishment—something we might simply term "culture"—behind him?[10] For one thing, if a refuge from other sites of culture, the family can simultaneously offer a more intense version of its pressures.[11] By supplying, as Gary Waller suggests, a borderline between the biological and the social, the family may nourish as much as bar historical forces or social demands.[12] At the countess's Wilton estate, literary circle and family retreat appear identical. But Philip Sidney's *Arcadia* and *The Defence of Poetry*, written there after he retired from Elizabeth's court, were also assiduous attempts to refashion home ("England, the mother of excellent minds" having "grown so hard a stepmother to poets"[13]). The problem is how to analyze family dynamics without dismantling them. Even if the Sidney family circle is not a fiction, according to Jonson's celebration in *Penshurst* it could be as ephemeral as one.

The sometimes-confusing web of positions and alliances and pieties marking the Sidney circle (which also included authors like Raleigh,

Spenser, and Greville) could later thin to protect—or quietly conceal—cousins Mary Wroth and William Herbert (Robert's daughter and the countess of Pembroke's son, respectively) when they became lovers.[14] But such familial ties were nonetheless both private and public, biological and social, carefully worn in some places, carelessly tangled in others. The literary efforts of the Sidneys also reflect this knotty design, as John Donne comments in his 1635 poem on the unappreciated "Sydnean Psalmes" coauthored by Philip and Mary, describing an alliance "[s]o well attyr'd abroad, so ill at home." When Donne tries to anatomize a part of this divinely corporate structure he also lets it slip through his fingers, claiming the authors of the *Psalmes* are at the same time: "Two, by their bloods, and by thy Spirit one."[15]

Philip's siblings' literary efforts continue to exemplify this divided, sometimes uncertain poetic, even as they try to overcome or repair it. Even Sir Henry's instructions on imitation belong to a larger family enterprise centered on self-promotion and humiliation. One of the most famous episodes in the Sidney family history took place at a tilt day exhibition for Elizabeth, where Philip's shield wittily featured the emblem "Speravi" (I have hoped) carefully crossed out, symbolizing the way his dynastic hopes had been dashed after Leicester produced an heir.[16] The fortunes of memory have their own props and symbols, and Philip makes himself their servant and arms bearer. This homely, somber task marks his pious exercises, too. Sidney's translation of Psalm 31, "In Te, Domine, Speravi," imagines the cancellation of memory by a poetics of repetition and recognition ("Now I, now I my self forgotten find"), "dreamed out of mind" when he finds himself dead.

*   *   *

For all of the disengagement Stone describes, however, many aristocratic Renaissance families effectively marshaled themselves in service of themselves. "Great families" like the Cecils, the Herberts, and the Cavendishes encouraged virtuous action or bloodshed, glorious achievement and vainglorious behavior, their stories "allegories of their time."[17] To a great extent, moreover, such sophisticated (if unsubtle) genealogical projects like the one written and rewritten across Sidney's arms were avidly pursued by many noble families, as Stone has carefully outlined, arguing that the frenzied buying of titles at the Tudor and Stuart courts spawned intense competition between a landed but decaying aristocracy and a rising merchant class.[18] The newly titled Sidney family may at first have profited from this practice, but older families often deteriorated:[19] evoking Sidney's defaced motto that makes war on ambition, Hamlet's dead father will return to the stage encased in armor after his brother murders him and steals his title. Even Prince Hamlet himself will decide to play a part in order to assert his own royal claim.

The "knowable community" Raymond Williams has described— once available at court or, at least, assumed like aristocratic rank and royal favor—became increasingly restricted and private in early modern England.[20] What would come to serve as a nineteenth-century *rural* ideal was at this time relocated to isolated family estates like those at Wilton or Penshurst, as Williams reminds us in a reading of Jonson's homage to Robert Sidney's country home.[21] But these centers of presumption and knowledge (where courtly fame was replaced by easy familiarity) could themselves be disruptive; Penshurst Place was constructed, after all, by evicting tenants and fencing in the magnificent surroundings, and Wilton had been the site of abbey lands confiscated by Henry VIII.

Great family centers had other claims to overturn and, like Penshurst's former tenants, perhaps Philip Sidney's image haunted the borders of his family circle. "[R]are ornament" and exiled poet, the adult Sidney often seems an outcast or refugee from courtly intrigue (even, sometimes, when he was at court).[22] Duncan-Jones reminds us of the literalness of his unhoused state, for before his short stint at Flushing, Sidney had stayed at his sister's home while Robert made a tour of Germany and, once married, briefly lived under his father-in-law's roof.[23] "Perfecte paterne" and holy ghost, Sidney's writings reflect the discomforts of his provisional status, his work finally packaged, carefully framed, and shrouded by secondhand rumors and pious emendation. As Clark Hulse has commented on the elaborate dissembling in *Astrophil and Stella*, the interpretive efforts of Sidney's readers only extend the secrecy—and suggestiveness—which so powerfully house and shadow him.[24] Like his cancelled shield, Philip Sidney's image remains both witty provocation and muffled threat.

Similar threats were discharged, however, in all Renaissance families. Rather than serve as a site of unequivocal understanding, the household was home to jockeying for positions and interpretations, the earliest setting for sex roles and power plays, personality, disappointment, and merriment. As E. Anthony Wrigley writes: "The pre-industrial family was to a greater or lesser degree the chief unit of reproduction, production, consumption, socialization, education, and in some contexts, religious observances and political action."[25] Yet Renaissance siblings always constituted a danger to this mechanism: daughters required expensive dowries and sons who were not first sons often lacked means and education. Carefully transmitted by his father, Philip Sidney's "lovyng" example indicates how siblings might represent both an enduring protection of and challenge to immature Renaissance selves.

This is also because, with exceptions like Mary Tudor's royal innovation, siblinghood is not subject to change, less open to interpretation than for instance paternity. In the controlled environment of Shakespeare's theater, these harder lines of descent easily become impenetrable. Siblings like Claudius and Hamlet senior or Regan and Goneril challenge autonomy and aggravate selfhood, sharing thrones and spouses (Stanley Cavell points to this indeterminability when he suggests the names "Goneril" and "Regan" are anagrams). Shakespeare's sibling pairs signal deeper hermeneutical doubts at work culture-wide not simply over establishing proper claims but over whether there are enough blessings to go around—whether the world can truly sustain families if it rewards ungrateful daughters or suborns forgetful sons.

* * *

Renaissance siblings represented political possibilities as well as moral ones. Marriageable second sons in aristocratic families frequently proved to be either a liability or a "trump card"; describing the difficulty of locating eligible bachelors for the widowed Mary Stuart, Antonia Fraser explains that a second royal son "might easily combine the disadvantages of both the foreign prince and an inferior subject."[26] If siblings could supply families with second chances or escape routes (so that Cordelia both instigates and solves her older sisters' plots) siblings also embody wider reservations a culture might hold about attractiveness and desert, nature and will, self-restraint and self-absorption, forever looming as potent reminders of weakened ties.

Still other familial complications are explained by a paucity of early modern terminology. Hamlet reflects on the deficiency when he takes his uncle to be "A little more than kin and less than kind." The *OED* lists no usage of the term "kinship" prior to the nineteenth century. This suggests there was little feeling surrounding familial relation before the early modern period: either one was somehow part of a family circle or excluded entirely from it, the connections absolute rather than continually discovered. Interestingly, though, there are more than fifteen variations on "kin" including "kinswoman" and "kindred"; and Shakespeare often uses "kin" to mean "akin to," indicating generalized likeness rather than direct relation (see *OED* 1.d).[27] The sweeping panorama of characters in a Dickens novel seems miraculous next to the claustrophobic gatherings in the *Henriad*. There appears little room for an extended set of variations, only for an array of doubles, some less, others more like.

Perhaps this inchoate web of familial attachments and perceptions explains why the "blundering" Henry Sidney meticulously diagrams his relation to both his sons, comparing affections, supplying a formula, and providing a "dyrectyon," as if the family should be used as a map rather than serve as a destination. After all, its wiring also had to support the brutal law of primogeniture, which made the first son's claim the only claim and set up older brothers as powerful rivals who taught fathers their duty.[28]

This cultural practice designed to conserve familial resources could erode families in other ways. Describing primogeniture as a "grim fact," Anthony Esler maintains that the "fundamental law" of inheritance "condemned many a younger son to social inferiority, and compelled him to win back by his own efforts the honor and authority his elder brother would simply inherit."[29] Primogeniture thus instituted a radically different model of achievement in male siblings. Esler also points out (in a study that uses Robert Sidney as an extended example without any mention of his poetry) that "a surprising number of the most successful aspiring minds were in fact younger brothers . . . Sir Walter Raleigh, Robert Cecil, and Francis Bacon."[30] What is particularly striking is that when Robert and his sister Mary write poetry, they are taking up one of their dead brother's "toys," pursuing something Philip avidly sought to hide. At the same time, their efforts conscientiously abide by the blundering Henry's instructions to follow their brother's "rare" "formular," although they ultimately interpret Philip's example in completely different ways. Perhaps the Renaissance "self-fashioning" Stephen Greenblatt outlines occurred when biological fashioning was short-circuited, for it never seems complete.[31]

* * *

Freud speculated that childhood fantasies center around a wish that the child be shielded from threats presented by the family itself, that more loving parents would release him from obscurity or impoverishment and rescue him from home. Renaissance texts often put these wishes in adult hands, for besides Henry's letter to his son, a number of stories (like those recounted in *Lear* and *Hamlet*) adjudicate—and sometimes disallow—familial claims. If Machiavelli's *Prince* leaves out any mention of siblings in his discussion of "Hereditary Monarchies" and "Mixed Monarchies," this is because the prince's interest always lies in conserving power, not in transmitting it. We find instead a "general rule": "[T]hat whoever is the cause of another become powerful, is ruined himself; for that power is produced by him either through craft or force; and both of these are suspected by the one who has been raised to power."[32] According to this formulation, familial operations operate like a zero-sum game, where no good deed or younger brother ever goes unpunished.

A more obvious early modern guide to familial resources was provided by the Hebrew Bible, a primer of affective vicissitudes and fortunes. The fatherly instructions we find there reflect an abiding concern with the role of siblings, and there is even a precedent for Henry's fatherly displacements, some pious logic to them. Indeed, the Old Testament—from which early Protestant readers carefully derived many sentiments and structures[33]—is a history profoundly anxious to organize families by realigning their responsibilities and sympathies. Many stories in the Bible reflect generational conflicts as hermeneutical puzzles and treat sibling rivalries as philosophical problems. Confusions over which brother to favor or which sacrifice to accept become terrifying and immense when the stakes involve limited blessings or many gods.

In her study of the origins and legacy of monotheism, Regina M. Schwartz argues that the "inexplicable" rejection first of Cain by God and then of Esau by Isaac suggests we ought to replace the term monotheism with *monolatry* (one blessing).[34] The chronicle of favors bestowed on chosen sons recounted in Genesis seems clearly aligned with the English custom of primogeniture, which limited the bulk of an inheritance to a single male child. Frequently in biblical accounts the devotion of only one child is needed or desired, a restriction animating the envy that later divides Joseph from his brothers. As Schwartz maintains, the founding moment in this founding text is not the Fall or some "shattered imaginary"[35] but the story of Cain and Abel, the curse of lineage. The first family is almost immediately divided against itself, the paradisal circle not exactly closed off, but broken from inside.

In a way, Sir Henry strives to repair this broken circle by shrinking the world's moral weight to the radiant image of his son, reducing the wealth of worldly knowledge to Philip's "vertues" and "accyons." Just as crucial, he disowns any personal claim on Robert in order to elevate his first son, setting before Robert his elder brother's priority. Many features of the Sidney myth are similarly generous (or similarly myopic), setting up the dead Philip as literary precursor in order to provide a space for his followers to "neatly queue up" behind him.[36] Such playfulness or maneuverability is probably at the heart of many family narratives, and the child's wish for more glittering, accommodating parents has its own literary genre according to Freud, who calls it the family romance.[37] But in Henry Sidney's version, the fairies actually get things right. They place the longed-for baby in the proper cradle, even though the delighted father finds himself undeserving of such a blessing. And in this story the remote giant is a gentle shepherd, a "lovyng brother" who has no peer. The "Sidney family romance," as Gary Waller has termed it, continues to reflect this polite misgiving about familial relations and deserts.

Readers like John Aubrey and Waller have focused on the peculiar wishes endemic to the Sidney family in order to fashion narratives about an incestuous union between Philip and Mary; with less strain, we might use them to understand the web of erotic and familial displacements that inhabit Robert's daughter Mary's literary romance, *The Countess of Montgomery's Urania* (1621), as I explore in chapter 4.[38] But here I focus more squarely on the initial problem that makes these stories seem necessary: Philip Sidney's success and failure, his literary availability and exile. His ambivalent model compels his siblings to take him up as teaching tool or heuristic, someone to learn from and learn about and then—forsaking their father's advice—push aside. The poems Philip's siblings produce are important texts in the story of what culture could take from such a figure and what it would leave behind.

*   *   *

It is Mary Sidney Herbert, the countess of Pembroke, who presides over her late brother's literary estate, overseeing publication of *The Defence* and *Astrophil and Stella* and a revised copy of the *Arcadia*, as well as completing her brother's psalm translations. But what did Mary, a writer more eminent than Robert (and probably, ultimately, more influential than either brother) learn from Philip?[39] And what did Robert make of his brother's literary example? If Mary, at least until recently, was more recognized as "Sidney's sister, Pembroke's mother," Robert has generated interest among critics mainly as Sidney's brother, Wroth's father. The Sidney family dynamics create a situation where text becomes context and back again, but I think these kinds of transformations are a feature of family life in general. Perhaps such dynamics are what enabled Renaissance families in particular to stay alive or to flourish, to turn threats into blessings and blessings into threats. This can also explain why what Robert takes up from his brother is so different from what Mary derives. If she exploits Philip's potent anonymity, Robert anatomizes the defeats and inconsistencies such anonymity often fostered.

Obviously, though, they are not the only writers left to determine what to take from Philip Sidney, nor was the problem of his legacy solely confined to writers. Queen Elizabeth shrewdly detected the way Sidney's legacy had been parceled out when she selected her favorites, severing the pleasing courtier from the dutiful servant so that, as J.B. James notes, "Elizabeth [later] found it hard to resist spoiling [Essex, the] young impetuous stepson of her adored Leicester, and easy to ignore [Robert Sidney,] the faraway, uncomplaining Governor of a Dutch town."[40] But Philip's siblings provide the starkest images of his influence, two pictures of Philip Sidney in relief. In examining Robert's and Mary's poetry, I want to establish how they believed their brother's different projects might continue, and what had to be emphasized or suppressed in order for this to occur.

Gary Waller claims that Mary succeeded because she had Philip's model before her, her portion of the *Psalmes* "written under her brother's (absent) tutelage."[41] But Robert was instructed to make use of his brother's example as well. Philip provided him with money and a reading list when Robert made his tour of Germany, and positions himself, albeit obliquely, as an audience in one letter, where he faults Robert's handwriting: "you write worse than I, and I write evil enough."[42] Perhaps Henry's fatherly injunction blocked Robert from the kind of success his sister Mary has come to enjoy, like the "covering cherub" or writer's block that Harold Bloom stations at the gate of literary paradise. Many of the myths surrounding Philip Sidney not only explain his incandescent presence but also suggest how and where Renaissance obscurity could reach, yet we can also take up the poetry of his siblings to locate some of these bounds. Of course, the limits of obscurity is a theme fledgling Renaissance writers frequently take up, nowhere more clearly than in Milton's *Paradise Lost* (a history of fatherly illumination and sibling disenfranchisement). *The Faerie Queene* offers another example of these limits in Spenser's Florimell, who evades most admirers and makes shoddy imitators disappear.[43]

*   *   *

The death of an author leaves behind an enormous imaginative burden, whatever liberties it permits writers and readers. This shadow self takes up room and craves attention; as Hamlet discovers, a ghost can speak more forcefully than any sleeping king. But what if this shadow self has different voices? Philip's unfinished, unsung pieties seemed to stimulate Mary on the one hand, while the ceaseless courtly machinations of a lover-artist encouraged Robert on the other. Such disparate lessons suggest we reorient our understanding of "influence" by returning to the term's etymological roots, seeing it as something that might work across familial lines rather than along them. After all, the term "influence" is tied to *influenza*, a contagion more likely to spread in especially close quarters.

This early modern sense of influence as something infectious or draining can be examined, too, in light of the simultaneous valorizing and degrading of Renaissance familial relationships that Stone chronicles. We see this at work in the dislodging of the Virgin Mary's divine status and in Henry VIII's many marriages, operating on a lesser scale in Henry Sidney's letter. Some of these familial alterations characterize the "forging" of English literary lineage at this time as well, which pushed aside the faraway Chaucer, disdained the more recent Skelton, and no longer relied on the Bible as a text on which to graft all later poetry. Continental models were increasingly favored, but a native "precursor" like Philip Sidney was needed too, to stand at the head of the English line.[44] Scholars like Raphael Falco, John Guillory, and Richard Helgerson have explored these early modern transformations of literary genealogy to trace "the history of authority," which, Guillory reminds us, is the "real history of authors."[45]

Falco describes massive efforts by English poets to "fabricate" Philip Sidney as their progenitor, efforts that worked so well because the exemplar was dead, his threat mitigated by loud and public mourning. Commenting elsewhere on the "flexibility of genealogical truth" at this time, Falco argues that models of literary paternity would do better to note "father-in-laws" instead of fathers since there was little that was natural or inevitable about this reproductive procedure.[46] Suggesting the same kind of genealogical reconstruction (and the same kind of heavy lifting), Guillory analyzes the disentangling of secular from sacred motives during the Renaissance, pushing God into the heavens where dead poets reside. This division encouraged Spenser's and Milton's reworking of biblical and mythical history by relocating authority from inspiration to the imagination. Finally, Helgerson describes how poetic debts could be discharged and influence buried when early English writers, plagued by anxieties about their "prodigal" art, passed off their productions as frivolous pursuits.[47] Poetic careers were motivated by careerism, and the "paterne" Sidney provides was "perfecte" because behind it could be located the bad faith or childish wishes and professional errors of a generation.[48]

Helgerson (1983) reminds us too that, unlike Spenser, who alerts the world to his promise, or Shakespeare, who earned money from his literary efforts, Sidney never profited from poetry or recommended it as a useful activity: instead, it provided consolation for real activity, ensuring a surplus of Cyruses in place of a real one. Indeed, it is this excess of models that provokes the poetry of Robert and Mary. To them Philip leaves two disparate images of the poet's calling to contemplate and finally dissever. Their lyric projects are elaborations and reconsiderations of Philip's metrical psalm translations and *Astrophil and Stella*, and in these projects they seek to perfect the pattern Sidney had only tentatively established.

Love poems were typically the province of male poets, and religious poems, especially translations, were one of the few areas where women writers might safely circulate their work, so the apparent differences between Robert's and Mary's efforts are not all that surprising. What is striking is that each sibling seems to have begun writing at a time when the dead Philip was only beginning to be regarded primarily as a poet. It moreover appears that both Robert and Mary were working on their respective projects at the same time. Mary likely completed her translations between 1597 and 1599, while Robert's work is typically dated between 1596 and 1598, the period of his longest stay in Flushing, where he assumed his brother's post as governor in 1588.[49]

The paths Sidney's brother and sister took seem irreconcilably opposed, as we will see in the following pages. But Gary Waller points to a contradiction nearly always at the heart of much of Philip's work, the divide between Protestantism and Petrarchism, "Calvinism" and "courtliness"—implying that the center would not hold for long anyway. Roland Greene locates a source for these pressures, however, in the even deeper fissures within the lyric tradition itself, exposed early on in Petrarch's *Canzoniere*, a series of "commemorative fictions" that signal the lover's inability to consummate his desire or repair the ruptures of his earthly, tainted vision.[50] Such fictions obsessively detail an idiosyncratic and often disillusioned private self at odds with convention, unreconciled to transcendental aims, and at war with his lady. They remake memory as the site of fragments and half-truths, hollow promises and crushed hopes, a place to collect one's shattered energies rather than connect them.

Both Philip Sidney's poetry and his actual if thwarted efforts as courtier and lover furnish models of the "commemorative fictions"

Greene describes. But to imitate Sidney would expose and subvert his example, making something flawed perfect, something halted articulate and whole. Moreover, Sidney wrote his own misunderstanding into the poetry, protecting himself from those effects by anticipating them. The first sonnet of *Astrophil and Stella* alludes to this cautiously ambitious program when Astrophil lays bare the divisions between poet and lover: "LOVING in truth, and fain in verse my love to show," he proclaims himself sincere because he is confused: "others' feet . . . but strangers in my way" (ll. 1, 11).[51]

In Sonnet 15, Astrophil excuses himself from the charge of merely copying Petrarch while preventing his own followers from easily emulating him. He addresses his rivals (dead poets as well as contemporary lovers) directly, his love poem at the same time a battle cry:

> You that do search for every purling spring
> Which from the ribs of old Parnassus flows;
> And every flower, not sweet perhaps, which grows
> Near thereabouts, into your poesy wring;
> You that do dictionary's method bring
> Into your rhymes, running in rattling rows;

This sestet mocks cloying versifiers who are foiled by lame rhymes and empty repetitions: for them "every purling spring" is a dirtied well and Petrarch's prized laurel wreath bent beyond recognition, their disfigured "poesy wring":

> You that poor Petrarch's long-deceased woes
> With new-born sighs and denizened wit do sing:
> You take wrong ways those far-fet helps be such
> As do bewray a want of inward touch:
> And sure at length stol'n goods do come to light.
> But if (both for your love and skill) your name
> You seek to nurse at fullest breasts of fame,
> Stella behold, and then begin to endite.

The speaker's expert uncovering of fraud ("stol'n goods" "come to light" once Petrarch's woes are dead), conducted like a criminal investigation or autopsy, uncovers its own aims, too, only to replace them with the tender comforts of the nursery, where the conscientious poet reposes at Stella's lap (with no want now of "inward touch"). These childish comforts were explored in chapter 1. But we also saw how fame and familiarity can merge, as when Astrophil rewrites genealogy and power, substituting literary tradition with mother's milk and "every purling spring" with Stella's breasts.[52]

*   *   *

It is stunning how the fictions of *Astrophil and Stella* so often yield to such satisfyingly speechless, even childish consolations, where youngsters easily triumph over the elders they will not imitate or ornate language surrenders to unintelligible marks. The muse finally tells Astrophil: "look in thy heart, and write." For Robert this will prove a hard model to follow. In contrast to the indulgences proposed by *Astrophil and Stella* are the sterner consolations of Philip's psalm translations, where regular or ritual satisfactions are suppressed in favor of the relief won through renunciation or distance.[53] Communal song is silenced in the Sidneian psalms by a radically austere image of the poetic self, isolated and uncomfortably adult, for this self is not whole, nor entirely knowable, nor always worthy of emulation.

Sidney's recasting of the psalmist is part of a larger Renaissance project that involved many singers and translators. As Greene comments: "In the sixteenth century the Psalter stood . . . for a gathering place of ritual and fictional impulses, an original harmonizing of these impulses and a challenge to translators—including some of the most accomplished poets of the age—to attempt such a harmony afresh in the new medium."[54] Greene suggests, though, that biblical poetics had long presented this challenge to translators and imitators, likely prompting responses such as Petrarch's in the first place: "Out of its Western sources in the first Psalm and the Song of Songs, the matter of scattered fragments [was] crucial to the fiction of the *Canzoniere*."[55]

Knowledge of this context helps us to put Mary's relative obscurity in better perspective, for if the completed psalms are among Mary's most accomplished efforts, her work (and its success) has not always been fully acknowledged.[56] Part of the uncertainty over Mary's authorship is further explained by the massive project of psalm translation itself, since with the Renaissance discovery that the Hebrew psalms were poetry, not prose, verse editions that could be sung in early Protestant churches enjoyed something of a vogue. Many translations quickly surfaced, their understated simplicity and directness (and the ballad form in which they typically appeared) perfect, Coverdale claimed, for the ploughman or spinner who could sing but not read, as well as for the prisoner or martyr who might voice them in protest.[57]

Philip's psalm translations seem to rework this "powerful and authoritative mass of verse"[58] with the aim to reach past poetry. He boldly experiments with meter (never resorting to the common meter so popular at the time) as if he aims to expose the failings of English verse outlined in his *Defence*.[59] We get a fuller sense of the "straw poesy" he derides there in the psalms he completed (numbers 1–43), where unpersuasive lovers are reimagined as uninspired blasphemers. These poems constantly muse on the speaker's innovations and isolation, melding grace with originality. Psalm 4 offers an account of the poet's genealogy that rejects the comforts of familiarity for the possibilities of exile. Fathers and children are dislodged (as they were in Sonnet 15 of *Astrophil and Stella*), replaced by the union of God and his chosen son, and Calvinist ambition is rewritten into a Protestant family romance:

> O men, whose fathers were but men,
> Till when will ye my honor high
> Stain with your blasphemies? till when
> Such pleasure take in vanity,
> And only hunt where lies do ly? (ll. 6–10)[60]

> Yet know this too, that God did take
> When he chose me, a godly one:
> Such one, I say, that when I make
> My cryeng plaintes to him alone,
> He will give good eare to my moane. (ll. 11–15)

This speaker's world has shrunk, its inhabitants and discourse drastically scaled back, but his repetitions magically ensure a kind of moral finality (the speaker castigates "men" "whose fathers were but men") so that God can only choose a "godly one." Glossolalia or pure sound becomes sanctified speech and the speaker's "moane" the best form of prayer.

Sidney's speaker describes a bond unavailable to anyone else through a poetry that cannot be heard, copied, or understood:

> O tremble then with awfull will:
> Sinne from all rule in you depose,
> Talk with your harts and yet be still:
> And when your chamber you do close,
> Your selves, yet to your selves disclose. (ll. 16–20)

Abandoning his enemies to privacy and prayer, Sidney's speaker suggests the worst fate is to be faced with self-knowledge in solitude. Not only, in contrast, does the speaker have an audience with God but God absorbed by culture in ways their older brother could not because of their strong misreadings and overwhelming affection for the dead.

Perhaps obscurity is something daughters and sisters rather than first or second sons can better explore or cultivate—it provides a language and a rationale, a comfort and a test. Perhaps for this reason, Mary Herbert's language has lived on, as other critics have noted, in Protestant linguistic habits of memory and community with an array of broken relics and missing muses at their disposal—Donne's bracelet, Wordsworth's Lucy Gray, Blake's Jerusalem—Philip's broken body only the first of many failed images that are accompanied by a range of enduring and contradictory interpretations.[70] To be sure, there were simpler ways for Renaissance culture to make family circles cohere. Mary Stuart's son James VI was ultimately recognized by Elizabeth as her heir, and James assumed the throne remarkably smoothly after Elizabeth's death in 1603, having kept his silence years before when his mother had been executed by her cousin. With the accession of James, Robert was finally recalled home to England, and in 1618 the king bestowed upon Robert his uncle Leicester's title, a reward Philip had long sought and never won. Robert Sidney seems to have finished out a long life by carefully presiding over his own celebrated estate, tending the fenced-in groves and fields at Penshurst Place.

PART II
Compromising Positions and Rival Romances

CHAPTER 3

THE GRAMMAR OF FAMILIES IN
SIDNEY'S OLD *ARCADIA*

Among Milton's many complaints in the divorce tracts (1643–44) is the shrill claim that in them he is writing "no meer amatorious novel" (*DDD* 2.256), a phrase elsewhere linked repeatedly to Sidney's *Arcadia*.[1] Yet Sidney's unusual picture of married life and domestic embarrassment in the Old *Arcadia* is strangely suggestive of the torpor Milton himself later outlines: the frustrated Gynecia and wearied Basilius who find themselves in bed as much resemble two sorry inhabitants of Plato's cave, stranded together and in the dark, as they do the graphic picture Milton will summon up sixty years later, "where no correspondence is of the mind . . . instead of beeing one flesh, they will be rather two carkasses chain'd unnaturally together; or as it may hap'n, a living soule bound to a dead corps" (*DDD* 16.326).[2]

My focus in this chapter is not upon Milton's romantic frustrations or his distaste for the genre of romance but upon the fruitful union of poetry and prose in Sidney's work. So unlike the bodies Milton describes, "chain'd unnaturally together," their union, I argue, characterizes Sidney's larger romance project, one that Milton himself takes up as well. I would suggest at the outset, however, that Milton's criticisms in the divorce tracts derive as much from the sour grapes of a literary heir as they do from any misery with Mary Powell, and that similarly private longings and discomforts are chronicled by Sidney throughout his prose works, as in the imagined account of Elizabeth's married life with Monsieur.[3] Simply put, the genre of romance is supplied a means for solving some of the domestic problems Sidney and Milton likewise depict when it can represent them in prose.

* * *

For both Sidney and Milton, prose was a useful instrument of state. Yet the politic Sidney was not known, either by his readers or by his admirers, as a master of indirection. Sidney's muse tells the blundering sonneteer Astrophil, for example, simply to look in his heart and write. And Sidney's bluntness often courts danger. In his tract defending Leicester, Sidney challenges the rumor mongering author of *Leicester's Commonwealth* (1584) to a duel, even while admitting—and exalting in—a lineage distinguished by murderers, traitors, and royal usurpers. Along similar lines, his 1579 letter to the queen protesting her planned marriage to Alencon brazenly tangles with the same delicate subject matter that cost John Stubbs his right hand. Sidney's Old *Arcadia*, a prose romance written shortly thereafter in retirement from Elizabeth's court (and later partly redrafted before his death in 1586) seems like an exception to an extraordinary career marked by frank poetry and rude prose.

Prose is typically esteemed (when it is esteemed) for its clarity and directness, but this is a modern prejudice, for Renaissance prose is more notable for muddled and exuberant verbiage. In Sidney's hands, prose becomes a "mixed mode" with the *Arcadia* its most artful example, a heady concoction of rhetorical exempla and poetic image, epic ambition and romance maneuvering, learned satire and gentle moralizing.[4] Sidney's prosody in the Old *Arcadia* is sensuous and logical, extravagant and precise, his methodology especially clear in the description of the sisters Pamela and Philoclea in Book 1:

> The fair Pamela, whose noble heart had long disdained to find the trust of her virtue reposed in the hands of a shepherd, had yet, to show an obedience, taken on a shepherdish apparel. . . . But when the ornament of the earth, young Philoclea, appeared in her nymphlike apparel, [she was] so near nakedness as one might well discern part of her perfections, and yet so apparelled as did show she kept the best store of her beauties to herself. (*OA* 33–34)[5]

Such a carefully planned artifice of evasions and exposures (where one sister balances and challenges the other, "virtue" clashing with "nakedness," abandoned disdain alongside unconscious self-interest) furthers a narrative that pulls apart conventional romance scenarios by unmasking unconventional characters, a narrative always straining to restore meaning to a romance universe fabricated out of costumes, reversals, poor readings, and opposing interpretations. If a "vanguard rhetoric," Sidney's prose is at the same time a conservative instrument too, highly cognizant of the epistemological and generic bounds with which Sidney plays.

Sidney's prose in the Old *Arcadia* is also a keen register of the manifold troubles and specious insults of family life. Leonard Tennenhouse comments that here Sidney "plotted out every permutation of Elizabethan desire";[6] but Sidney painstakingly plots out other forces as well: the political strife dividing parents from their progeny; the pastoral temptations that transform elders into children; and the profound disappointment and confusion that can befall nations when unseemly courtships are initiated.

Beyond these explorations into literary and political regimens (or habits), Sidney's prose plumbs psychic regions buried by powerful adult reason, returning to the haunts and cares and blurry lineaments of family life. In the Old *Arcadia*, his prose wrestles with the threats of an "uncouth love" and adultery (*OA* 5), all the while trying to solve the problem of exogamy.[7] These familial complications are predicted by an oracle that the duke Basilius consults at the very opening of the story:

> Thy elder care shall from thy careful face
> By princely mean be stolen and yet not lost;
> Thy younger shall with nature's bliss embrace
> An uncouth love, which nature hate the most.
> Thou with thy wife adult'ry shalt commit,
> And in thy throne a foreign state shall sit. (*OA* 5)

Romance will be turned inside out in the Old *Arcadia* in order to fulfil its promises, Sidney's story cataloguing and untangling multiple cases of "love against love" (*OA* 61). Romance confusions plague kings and stewards alike, and even the harebrained Dametas finds himself admonishing his daughter at one point: "Dost thou not know thy father? How hast thou forgotten thyself?" (*OA* 232). This difficulty in making out the faces and hearts of loved ones is fully redressed by Sidney's prose, however, and at the end of his story, the heroics—or what Basilius's counselor Philanax calls the "naughtiness" (*OA* 332)—of two young princes is rewarded by romance, although its conventions have now been exposed in terms of a fragile marital bliss, the replacement of cruel parents with disobedient children, and the claustrophobia of an entirely familiar domestic world.

Sidney's prose works repeatedly concern themselves with the poetics of exogamy, the project to preserve valuable lineages by twisting or supplanting them and strengthen social claims by attenuating biological ones—a project undertaken with the aim of making the world at large one's home, or making romance out of everyday life. Shakespeare's *Lear* shares this aim, but with tragic results: if a daughter's marriage is precipitated by a father's wish to retire to her nursery, her coming into adulthood ultimately deprives them both of power. As Christopher Martin maintains, "constriction to and within the circle of domestic immediacy [threatens] social chaos and collapse."[8] Gynecia's lament in the Old *Arcadia* lays bare these dangerous inversions of affection and authority when she admits her love for her daughter's suitor: "[I]t is my daughter which I have borne to supplant me," she confesses. "But if it be so, the life I have given thee, ungrateful Philoclea, I will sooner with these hands bereave thee of than my birth shall glory she hath bereaved me of my desires" (*OA* 81).

As I have already suggested, similar inversions shaped many early modern families, anxious to recognize themselves and their progeny in an increasingly unfamiliar social and economic universe. The prose taking shape during this time in cony-catching narratives, cheap pamphlets, and popular romances is a particularly fine instrument for the delicate maneuvering demanded in such a world because it allows the rude details of home life to be reconstructed as a means of advancing romance fantasies and projects—projects that Philanax, again crudely, reduces to the killing of a father and ravishing of the child (*OA* 337). Arcadian cousins become insurrectionists, parents and children erotic rivals, and sisters criminals. As I describe more fully later, Sidney reconstructs home life similarly in his letter to the queen. Both prose works are marked by an economy of sentiment and surfeit of stories that reconfigure families and their values. This new picture of family life—its harsh obligations and inequities and narrow consolations—is articulated by Basilius who, reconciled at last to Gynecia, announces: "so long as he unworthy of her did love, she would be the furthest and only limit of his affection" (*OA* 241).

*　*　*

Economy and excess are also important features of Pyrocles's disguise. The prince's metamorphosis into the fair maiden Cleophila is anything but complete, as he confesses to his beloved Philoclea when he discloses himself to her in Book 2 (*OA* 105). He tells the princess: "Behold here before your eyes Pyrocles, prince of Macedon, whom you only have brought to this fall of fortune and unused metamorphosis; whom you only have made neglect his country, forget his father, and lastly forsake himself!" In this admission Pyrocles points to the reversibility of his condition along with its interrupted or untapped state, but if his metamorphosis is something "unused" or untried it is also redundant. For one thing, Pyrocles's "womanish apparel" is nearly everyday equipment for the prince, at the most, something effortlessly fabricated, his transvestism the narrator implies really an emblem of courtly *sprezzatura*:

> [T]o begin with his head, thus was he dressed: his hair (which the young men of Greece ware very long, accounting them most beautiful that had that in fairest quantity) lay upon the upper part of his forehead in locks, some curled and some, as it were, forgotten, with such a careless care, and with an art so hiding art, that it seemed he would lay them for a paragon whether nature simply, or nature helped by cunning, be the more excellent. (*OA* 24)

In fact, Cleophila's appearance is superficially no different than Pyrocles's: the jewels and fabulous clothes are the accoutrements of a prince; even his coiffure is less brazen artifice, more golden nature. The fiction simply maximizes the effects of Pyrocles's beauty, the calculated artifice a more persuasive form of rhetoric or more provocative argument, "like a diamond set in a more advantageous sort" (*OA* 25).[9]

One might argue, however, that even this slight change of signifying practice is "an attack on authenticity and authority,"[10] and that other metamorphoses in Sidney's story are other subversions of nature or order. Philanax, perhaps predictably, describes Pyrocles's image as one of "whorish beauty," "the highway of his wickedness" (*OA* 338). But Arcadian metamorphoses can be quite subtle, and Sidney's attacks have many targets. If Pyrocles turned Cleophila is Sidney's wry projection of himself as sometime-prince and object of devotion to men and women alike, then the rustic Dametas is another projection, this one of a clown, skeptic, and poor friend to princes, "ill gartered for a courtlike carelessness; only well shod for his father's sake" (*OA* 27).[11] I say this in part because only the clueless Dametas seems to register something about Pyrocles that all of Cleophila's fawning admirers cannot. "[S]tanding upon his tiptoes, and staring [at Pyrocles/Cleophila] if he would have a mote pulled out of his eye, 'why,' " exclaims the puzzled Dametas to the Amazon, " 'thou woman or boy, or both, or whatsoever thou be' " (*OA* 29).

Because Cleophila's disguise is *almost* transparent, it permits Dametas to construe a third narrative possibility or "mixed mode," "woman or boy, or *both*." Perhaps Dametas's awkward formulation is encouraged by the prince's thin disguise, but such a clumsy epistemology also helps one comprehend Sidney's elaborate narrative discourse in the *Arcadia*. On the "margins of otherness" like Pyrocles in drag, it is a language betwixt and between, alternately poetry and prose, and, sometimes, both.[12]

If Pyrocles's altered state supplies a useful model for the contradictory shape of Sidney's career, Sidney's language in the Old *Arcadia* and in other prose works is another key to the equivocal nature of Sidney's persona, alternately appealing or awkward, careful or careless, untroubled or unquiet. Many of Sidney's prose experiments undergird experiments with his own vexed identity—with the displaced syntax and conflicting grammar of aristocratic life, clashing ethical codes, drastic economic change, and competing audiences for his efforts. Arcadian prose provides a haven for pastoral sentiment and a challenge to it; Arcadian poetry rewards prose entanglements with solutions or brief escapes.[13] Such disparate ways of grasping reality are also discursive structures shaping it, and in the hybrid *Arcadia* they are given flesh and costumes too, manipulated as a way of establishing Sidney's morphology as poet.[14]

Metamorphoses of character and syntax and grammar may also act, according to Leonard Barkan, as a figure for all of the fears and necessities of exogamy,[15] a movement across classes or countries that threatens the charmed circles of childhood and the family. This movement requires seeing oneself as an object of negotiation of transformation, as a work in progress. Sidney's prose reflects tremendous uncertainty over how one should conduct himself in such a world, where one must possess "multiple literacies," the ability to read and be read on different levels.[16] This problem is given shape in the disguise of Musidorus, Pyrocles's cousin, which Musidorus adopts in an effort to persuade Philoclea's sister Pamela that he is worthy of her. Since Pamela has been placed under Dametas's care, the rude speech and clothing of an Arcadian shepherd named Dorus gives Musidorus the latitude to address her indirectly (*OA* 87–90).[17]

Neither obvious prince nor true shepherd, Musidorus has no real claim on Pamela, yet the amorous rhetoric he uses to court Mopsa, Dametas's obtuse daughter, expertly cuts through Pamela's defenses (*OA* 92–93). In fact, Pamela soon comes to believe the handsome shepherd exactly because she knows he's lying:

> [T]he more [Pamela] marked the expressing of Dorus's affection towards Mopsa, the more she thought she found such phrases applied to Mopsa as must needs argue either great ignorance or a second meaning in Dorus; and so to this scanning of him she was now content to fall, whom before she was resolved to banish from her thoughts. (*OA* 87; see also 88–94)

Joel Altman views Musidorus's narrative strategy as a mirror for Sidney's:

> Musidorus recounts, in the guise of another man's adventure, the story of his own education, travels, love-sickness, transformation into a shepherd, and courtship of the King's elder daughter. It is not only a witty way to identify himself, but also demonstrates in small Sidney's fictional method,

for Pamela has been offered a clarifying lens through which she may learn how a noble prince, educated in a great court, might happen to fall in love and be forced to assume an unworthy appearance before his lady.[18]

But Sidney's subject is both autobiographical and aesthetic. His calculated prosody, its skillful excesses and clever confusions, instructs its audience to read between the lines, to relax syntactical constraints while more closely attending to rhetorical codes. Making stories out of the epistemological dilemmas inherent in storytelling,[19] Sidney's prose employs convention to analyze it, turning to its codes to cloak rather than to guide. As Dorus tells Pamela how to read his pastoral disguise, "this estate is not always to be rejected, since under that veil there may be hidden things to be esteemed" (*OA* 93).[20]

Like the characters of Pyrocles and Musidorus, Sidney's prose assumes different syntactic shapes and symptoms, variously embracing and retreating from the world of the poet.[21] In the process, the laws of the familiar world are rewritten and endogamy made something strange. Even the chaste Philoclea is alert to this transformation of the domestic sphere, "sometimes compar[ing] the love she bare to Cleophila with the natural goodwill she bare to her sister; but she perceived it had another kind of working" (*OA* 86).

There are other Arcadian inversions and reversals both affective and generic, so that the stoic Musidorus is quickly brought low before a sympathetic Pyrocles wearing a dress. Roland Greene explains the reasons behind such literary transformations of the familiar surface of things. He describes prose as operating through modes of "immanence" (which allow the initiated into a narrow moral or spiritual circle) or "embassy" (which merely permit visitors temporary access into a foreign world). Working through both modes, prose might provide a mechanism for both alleviating and untangling romance terrors, including those surrounding exogamy.[22] Such a shift preserves a genre otherwise on the verge of collapse. There are still other successes. It is his reliance upon both modes in his letter to Elizabeth that allows Sidney to take a hard look at his aging queen, tell her about her private longings, and then—with the nation's best interests at heart—counsel her against "so hazardous an adventure" ("A Letter" 47).[23]

*   *   *

Sidney's prosody seems equally sensitive to its author's secret longings and frustrations, a skill difficult for many of Sidney's literary descendants to duplicate: the poet's contemporaries were often as loving but seldom as penetrating as Pamela. Yet one of the first tasks his literary descendants needed to take up if they were to acquire a usable pattern or history was establishing exactly what it meant to assume a poetic career, and thus deciphering (or metamorphosing) Sidney's muddled legacy as poet and statesman, royal advisor, and aristocratic outcast. Theodore Spencer regards Sidney's legacy as enormously fruitful, if not completely obvious. "Sidney's attempt," Spencer claims, "whether invariably successful or not, is the significant thing, his attempt to be himself, to find a richer and more exacting freedom than that given by the *persona* of convention, the attempt that was to be Sidney's legacy, not only to the generation of poets which immediately followed him, but to all poets since."[24]

The problem is that Sidney's literary aim was to capitalize on his own marginal status in Elizabeth's court. Even Spencer's cautious praise suggests just how hard it would therefore be for the poet's literary descendants to emulate him: "To find his own voice," Spencer argues, "to discover his own poetic idiom and his own rhythm, is the main business of a poet."[25] What would it mean to take as one's progenitor someone who had disavowed poetry, construing it as a "toy" or idle pastime, something only undertaken in lieu of military exploits or a more distinguished role as one of the queen's advisors?

Sidney explored these questions himself throughout his brief career. His elaborate prose rationale for poetry outlined in *A Defence of Poetry* also constitutes a severe indictment of it, as if Musidorus's ambivalent positioning had been codified and celebrated. Sidney argues that composing poetry is a form of public service; at the same time, if a medicine for an "infected will" (*DP* 25) and cure for "naughtiness" (*DP* 31), poetry provides a way to seduce a dim-witted mistress (see *DP* 69–70), someone like the obtuse Mopsa, with her "special grace" (*OA* 88) and "virtues strange" (*OA* 27). In fact, poetry, being abused "by the reason of his sweet charming force . . . can do more hurt than any other army of words" (*DP* 55).[26]

Indeed, it is the splintered shape of Pyrocles that is really figured by Sidney's pronouncements in the *Defence*, where the poet's logic is obliquely and deliberately split in two. If readers of the *Arcadia* are frequently disturbed by the contradictions between the cousins' heroic values and their ignoble behavior, Sidney's example as a poet is similarly problematic, seemingly self-defeating. Following it means choosing to disavow the poetry one writes and calling into question the very idea of a poetic career. Yet such concerns about one's identity and how to enforce it are at the center of the Old *Arcadia*, given shape as amorous intrigue, peasant revolt, and pastoral conflict. All of these Arcadian crises revolve around a deeper question—both epistemological and ontological, both theoretical and physical—of how others should see us, a problem Pyrocles introduces with some embarrassment when he reveals himself to Philoclea as neglectful patriot, ungrateful son, and absent prince (*OA* 105).[27]

* * *

Roland Greene points to a similar crisis when he describes the "seeming imperative of early modern prose, to discover and address its own protocols of representation[,] often irrespective of genre or convention."[28] This hardly seems like an imperative either anxiously or carefully heeded, however. Renaissance prose was the most accessible, flexible, and disposable of literary genres, a vehicle for sermons, satires, treatises, romances, and for poets, as well. Many of the prose writers were, "of course," poets, as Kenneth Muir reminds us, "and they did not cease to be poets when they wrote prose. . . . There was none of the feeling of later periods that poetic prose was in bad taste."[29] Although the word "prose" nowadays seems to carry little meaning,[30] we would thus sell Renaissance prose short if we viewed it as a negative literary state, anti-imaginative or un-rhetorical, transparent and natural, "undefined, untheorized,"[31] or devoid of metaphor and opposed to poetry.

Ad hoc prose forms were "everywhere" in the sixteenth and seventeenth centuries,[32] as artful and as indebted as Renaissance poetry to classical and continental traditions. One critic suggests: "almost all the prose writers of [the] time, set [themselves] deliberately to produce a kind of artistic composition, following in some measure the accepted principles of rhetoric, in which the effect would be heightened by all the well-known devices of rhetoric."[33] Ann E. Imbrie adds to the picture of a rich prose tradition:

> the early humanists . . . had so revered prose and so relegated poetry to the leisure hours of life that by the 1580s—even without Stephen Gosson—poor poetry needed a good defense. . . . But in seeking to correct an imbalance in the literary values of their fathers, the sons among Sidney's generation incorporated rather than repudiated those values, defending poetry without necessarily maligning prose.[34]

The medium of prose accordingly presented Sidney with the opportunity to defend and simultaneously bracket his poetry. This benefit was not however limited to Sidney. From the first, prose's absorptive, synthetic qualities had made room for the mixed mode of *versiprosa* found in medieval romance, where it alternated with verse.[35] Sidney outlines an even more powerful fusion in *A Defence of Poetry*, claiming that "it is not rhyming and versing that maketh a poet" (*DP* 27). He goes on to cite the prose romances of Heliodorus and Xenophon as poems, and celebrates the "mingled prose and verse" of Sannazzaro and Boethius (*DP* 27, 43).

Sidney's reliance upon a prose and poetry that are interdependent isn't always obvious, especially in the *Arcadia*, where it appears instead that he divides the two. Maria Prendergast locates their apparent opposition in the distinctions between masculinity, reason, and didacticism on the one hand, and femininity, passion, and poetry on the other, Sidney's narrative continually shifting, in her words, from "agonistic deeds" to "stychomythic words."[36] But Sidney's efforts at securing literary boundaries and clarifying generic codes are repeatedly undermined or compromised by the narrative of the *Arcadia*. The wedding song of Kala and Lalus in the Third Eclogues, for instance, celebrates the consummation of Pyrocles and Philoclea rendered in prose in Book 3. As the narrator explains at the outset of the Third Eclogues, "I think it shall not be impertinent to remember a little our shepherds while the other greater persons are either sleeping or otherwise occupied" (*OA* 212). More thoroughgoing still, Philisides's formal laments are not only duplicated but solved by the romance questing of Pyrocles and Musidorus. Indeed, Cleophila makes the logic of prose apparent in poetry, singing: "What can justice avail to a man that tells not his own case?" (*OA* 74).

Obviously Sidney's prose has little traffic with what critics regard as the "realism" of the early novel, since it is equally ironic about and skeptical of empirical evidence and romance fantasy; but as I explore more fully in chapter 4, the early novel is a sympathetic environment for the emerging Renaissance family, of which Sidney's psychological perceptions are shrewd and formative. Actually, the poetry of the *Arcadia* is equally symptomatic of the new forms of affection and relation taking shape at this time. Cleophila ruefully complains in the Second Eclogues, for instance: "My muse what ails this ardour / To blaze my only secrets? / Alas, it is no glory / To sing my own decayed state. / Alas, it is no comfort / To speak without an answer" (*OA* 143). If Sidney's worlds of poetry and prose are divided from each other, at the same time they are cut off from all other discourse available to them. "[W]ise discourse, sweet tunes, or poet's fiction," as the shepherd's catalog runs (*OA* 132), each of Sidney's modes in the *Arcadia* renders other discourses equally fictive.

The Arcadian shepherds are likewise aware that pastoral rules have changed. One of them, the shepherd Boulon, describes these new patterns of relation in the Second Eclogues as a "grammar" through which one might "learn of more congruities" (132). Jeffrey Kittay and Wlad Godzich help explain Sidney's prose project of establishing new congruities, new patterns of meaning and relation. They help us see how the new signifying practices of prose are much like the ones arising out of the emergence of the early modern family, privileging nearness over identity, metonymy over metaphor.[37] Both rely upon a new form of interiority characterized by a self experienced in privacy, where privacy is defined as something adjoining, shaped by, and confined to a public realm, as something on display, in other words, at home. Although the sonnets of Wyatt, Sidney, and Shakespeare would suggest that the lyric sets aside a space for the interior self, this is a self at odds with the world. Prose narratives take another approach, illustrating how an individual belongs in and to the world as a piece of its prosaic and material texture, as a mouth to be fed, a desire to be quashed or met, a body and mind related to the thoughts and needs of others.[38]

As Kittay and Godzich put it, "prose is not an ingredient," since "it is not identical to any discourse or signifying practice it contains." "Among all the discourses it contains," they claim, "it takes the position that it is just holding them together."[39] Prose's poetics equivocate.[40] Yet the holding act is not without risks or casualties. In fact, Sidney's prose records the decay of patrimony and poetry in the same breath when we are presented with the image of the shepherd Dicus in the First Eclogues. Dicus bears a whip in one hand and a naked cupid in the other, along with the scandalous invective that "they had done the heavens wrong to make Cupid a god, and much more to the fair Venus to call him her son—indeed, the bastard of false Argus, who, having the charge of the deflowered Io (what time she was a cow), had traitorously in that shape begot him of her" (*OA* 57–58).

*　*　*

Of course, there was a larger background to the scandal Dicus relates. Constance C. Relihan accounts for the "explosion of prose" in the latter half of the sixteenth century in developments like the breakdown of the church and the decentering of Europe with the rise of travel and commerce.[41] Less favorably, John Carey explains how prose emerges from the "ruins of [the Renaissance] tradition of knowledge":

> The Elizabethans, living in the break-up of a culture, with a new religion, a new statecraft, a new social system, and new science at the door,

naturally enjoyed fitting well-worn materials into new patterns. It was a substitute for technology, and gave the same feeling of control. Experience is reduced to adage, thought to familiar manipulation.[42]

Along the same lines, M.M. Bahktin argues that novelistic discourse requires the destruction of mythological thinking, indeed, that this is its primary project.[43] But one might instead propose that Renaissance myth is being fed new images like that of Dicus, or being forced now to contemplate itself, the way Pyrocles sees himself in Cleophila as failed prince and bogus maiden. It is in this way that prose can extend without breaking the "charmed circle" of the Arcadian paradise, where "saying [can] make them believe whom seeing cannot persuade," where "outward utterance can command a conceit" (OA 92). A contemporary analogue of this shift in power (or shift in how power was imaged) is supplied in the aristocratic jockeying for royal favors and in the bogus genealogies noble families supplied to bolster their credentials.[44] When romance becomes self-conscious, more reflective about its fictive origins and methods, the worlds of action and meaning become increasingly horizontal as prose defines "modes of association" rather than "ontological states of being,"[45] new structures of filiation to replace outworn structures of power.[46]

Intelligence gathering and early modern diplomacy require such new modes, along with prose's speed and seemingly anonymous quality or disembodied state.[47] But prose also emerges when the early modern family is redefining itself and new discourses of relatedness are being coaxed into existence. We see these new modes of relation at work again and again in Sidney's writings. In his letter to Elizabeth, Sidney reminds her of her real place in men's hearts and in their stories.[48] Not only does Sidney use prose to make the queen a site of childish hopes and fears; through it, he also draws limits on the power any royal might wield. "Let not the scum of . . . vile minds" he cautions her, "Bear any witness against your subjects' devotion" ("A Letter" 56).

These prose moves are daring, but they belong to a broader reconstruction of authority in early modern England. Maria Prendergast describes prose in terms of a "transgressive aesthetics" "predicated on deviating from parental, political, and literary authority."[49] Kittay and Godzich portray prose as more of an orphan than a rebel, "as a signifying practice [that] has no corresponding accrediting agency. It has no fount of authority outside itself on which to draw and thus authorize the text that constitutes all its practice."[50] Throughout his career Sidney represents himself as bereft of such authority, sometimes at liberty as a result, at other times almost at sea. Prose gives him the power to advise royals, scold lovers, and explain his own ill-treatment, even allowing him all of these behaviors as when his surrogate Cleophila corrects the doting Basilius: "Your words, mighty prince, were unfit either for you to speak or me to hear; but yet the large testimony I see of your affection makes me willing to suppress a great number of errors" (*OA* 101).

Unfortunately, Sidney's prosody in the *Arcadia* has precluded much investigation of its impact on the novel. For one thing, prose isn't supposed to have an agenda. The general editor of *The Pelican Book of English Prose* explains, for instance, the "working-hypothesis" of the volume is that the prose collected there "should not be too self-conscious, that the writers of the best English prose usually had more on their minds than the problems of style."[51] Kenneth Muir turns to chronology for a similar rationale for prose's limitations. He notes that Tudor prose was often "syntactically immature, and it was not a good medium for exact description or subtle argument."[52] Franco Morenco observes in the *Arcadia*'s complicated prose design sharp moral fault-lines, claiming that "irruptions succeed each other with uniform consistency: they open a series of unhealthy parentheses in a state of normality which looks more and more precarious, until they lead to a final breakdown in the private life of the characters as well as in the life of the state."[53]

But there is another way to read Sidney's sentimental stories of lost sentiment, his prose record of failed poetry. We might understand Morenco's "unhealthy" parentheses in terms of a widening network of allegiances and contingencies, a picture of a vastly interrelated romance world in which congruities are as crucial as identity. When Sidney employs such fictions about fictions, he also tests his own position, the extent of his own metamorphosis, and the degree to which he can continue to play at outraged aristocrat and loyal civil servant, devout Protestant and worldly warrior, poet and nonpoet, favorite son and prodigal child. The thinning line between poetry and prose gives Sidney a fence to straddle as well as access into a set of ready meanings and easy stances that are rapidly slipping away.

*   *   *

Elsewhere, Sidney uses prose to expose the faulty claims behind (or thin walls separating) other discourses, to delineate the region John Pietro Pugliano inhabits in *A Defence of Poetry*, a zone of "strong affection and weak arguments" (*DP* 17). In this apology for his "unelected vocation,"

Sidney makes his uneasy position both subject and vehicle for a treatise on prose as poetry. As Roland Greene claims:

> Sidney's manifesto for poetry has often been read as though its value depends on its correspondences to his lyric sequence *Astrophil and Stella* and the two prose fictions called *Arcadia*, when it might be better understood as merely one installment in the both fictional and theoretical project in which all four texts are engaged, which can be defined loosely as a description of the borders between fiction and reality. As a work of first-person prose in his own voice, the *Apologie* might even be understood as a hybrid of the others—a third way of getting at the questions of literary ontology, epistemology, and ethics with which Sidney is preoccupied.[54]

The same epistemological shifting and ontological repositioning is on display in *The Defence of the Earl of Leicester*, where Sidney employs prose to turn the rumors about his uncle as traitor into a celebration of the Dudley family and transform his family's dubious legacy into a history of peerless peers, "no nobler" in England (134, 135). These prose reversals continually rewrite as they celebrate family history. At one point, Sidney exclaims, "So I think it would seem as great news as if they came from the Indies, that he who by right of blood, and so accepted . . . should be doubted to be a gentleman."[55] Later, as if to suggest that poor fiction derives from a failure to fully grasp familial relations, Sidney argues:

> For if either the house of Dudley had been great anciently and now extinguished, or now great and not continued from old time, or that they had been (unentitled) gentlemen, so as men must not needs have taken knowledge of them, yet there might have been cast some veil over his untruth. But in a house now nobel, long since noble, with a nobility never interrupted, seated in a place which they have each father and each son continually owned, what should be said but that this fellow desires to be known suitable: having an untrue ear, he will become it with an untrue tongue.[56]

Godzich and Kittay maintain that the writer of prose must be prepared to constantly disown his position. They comment that "[w]hat is required is the ability to abstract the *function* from the *agent* of communication, a process of abstraction and analysis [linked] to prose literacy."[57] Such a requirement seems unmet in Leicester's defense; in any case, what I am proposing has more to do with the ability of the agent, in this case Sidney, to make himself an abstraction, to metamorphose—like Pyrocles or Musidorus—but not completely, to be both loyal subject and rival prince, wishful thinker and vengeful knight.

Five years earlier (and coincident with the drafting of the Old *Arcadia*), Sidney had written to the queen, outlining a number of arguments against her marriage.[58] Here, the poet presents a novelistic narrative of failed romance, with an unhealthy knight, interrupted quest, and still-born offspring ("A Letter" 50, 57).[59] Indeed, he says that Elizabeth should remain single in order to continue to provide a "mirror" to her posterity (57), her monarchy so "perfect" because "so lineal" (54). However, Sidney's first move in this catalog of romance conventions gone dry or stale is to align the queen's body and happiness with his own. He promises "in simple and direct terms . . . [to] set down the overflowing of my mind in this most important matter: importing, as I think, the continuance of your safety, and as I know, the joys of my life" ("A Letter" 46). Prose then supplies Sidney with the means to picture the ideal marriage as a metamorphosis that would expel foreign elements and magically render things, like Pyrocles's cross-dressed beauty, only more themselves.

*   *   *

If a "dreadful expectation" sets in motion the plot of the *Arcadia*, forcing Basilius and his family into hiding, great expectations that bolster "so lineal a monarchy" finally right all the romance fears of incest and rape at its end. But, along the way, many familial ties have been thinned or confused or temporarily overturned. Miso comes to believe that her husband Dametas is cheating on her with their daughter, Euarchus sentences his son and nephew to death, and Gynecia's jealousy of her daughter turns murderous. It is prose that emerges out of such affective dislocations, as Godzich and Kittay argue, pointing to the dissolution of the older, local oral world in the twelfth century:

> The integration of material of such diverse provenance is difficult to effect, and new forms must be sought. What will be called upon is the same mechanism by means of which clan and family boundaries are rene-gotiated, namely a procedure of extension—the system of marriage and affiliation alliances. Thus are the great medieval narrative cycles born, one epic extending into another, massive works of a multiplicity of wandering knights. The eponymous heroes of these cycles turn out to be members of a vast kinship structure, one that can even be invoked to account for the undercurrents of tension that may exist among them.[60]

These "procedures of extension" allow romance to reconstitute itself, to identify strangers and extend familiar relations. They operate in

Sidney's letter to Elizabeth, where he imagines their affiliation as more crucial and more rewarding than the marriage of two royals. Another "procedure of extension" is what Musidorus envisions when he and Pyrocles adopt a second set of pseudonyms after their scheme to marry the princesses is discovered. Anonymity guarantees posterity in Arcadia: "[M]y virtuous mother," Musidorus tells Pamela, "shall not know her son's violent death hid under the fame will go of Palladius. But as long as her years (now of good number) be counted among the living, she may joy herself with some possibility of my return" (OA 271–72). Poetry would elegize the virtuous mother's son, but prose can disguise the prodigal. And families can be preserved through "procedures of extension" because these fictions are so closely bound with childhood fantasies of escaping one's home for a better, safer version of it, one that will never have to be relinquished.

The potion Gynecia inadvertently gives Basilius in Book 4 of the Old Arcadia characterizes a "procedure of extension" as well. First supplied long ago in order to guarantee Gynecia her bridegroom's love, then hidden away for years, the potion is recovered so that Gynecia might win her daughter's lover's affections; ultimately, the potion is the catalyst for a romance experiment that leaves things exactly as it found them in Arcadia, established marriages secure with new ones confidently arranged. Still, less interesting than how the potion actually works—its soporific effects are intensified by Basilius's advanced age—is how the mechanism of prose operates, the machinery Sidney employs to make a failed aphrodisiac yield such powerful regenerative effects.

Sidney's interest in providing Gynecia with a magical, exotic potion goes hand in hand with his desire to provide her with a history and psychology. Yet he doesn't stop there, offering an analogue for her behavior in her husband's own longings. His prose is generous enough to allow that his characters' minds—like the potion—do not work in a linear fashion or through a plain style but are ensnared by sympathies, misperceptions, lapses, and habits, all of which are the building blocks of family life.[61] Sidney's prosody actually searches out extenuating circumstances. It is not cluttered by mere ornament or euphuistic exfoliation, where ideas spring up (and fall away) like leaves, but is marked by a studied method of insinuation and implication. Sidney's romance is supported by "common disposition," the pragmatic economy and psychic routine of his Arcadian king and queen.[62] When Basilius miraculously awakens long after the prescribed thirty hours, he comes to his senses and praises his wife as a model of fidelity. Sidney's "medicine of cherries" (DP 41) finally drains romance fictions to preserve them.[63]

Gynecia's love potion's effects are analogous to Sidney's prose methodology. Both are techniques for shoring up a "shattered imaginary,"[64] that "imaginary," for Sidney, at once the household, a royal caress, and the poetry of such embraces. What Richard McCoy describes as a long-held "pattern of evasion and obscurity"[65]—movements in and out of the shadow of his family—is the fabric of this fiction Sidney weaves about himself, the gorgeous poetry of Sidney's prose.

CHAPTER 4

"MORE LIKE RUNNE-AWAIES, THEN PRINCES"

The genre of romance often proves a strained medium for magic, with its outlandish devices for rebirth or reform as well as the stock tools Samuel Johnson once faulted, "a hermit and a wood, a battle and a shipwreck."[1] For all of its heavy-handedness, however, romance is also a "persistent" genre:[2] it animates dream visions, pervades epic ambition, and structures many of the problems and solutions of the novel. It does all of these things, moreover, in the course of giving families their grand histories and precious meanings. When she began the romance of *The Countess of Montgomery's Urania* more than twenty years after the publication of her uncle's *Arcadia*, Lady Mary Wroth made use of the genre in order to recast the story of her life and her origins, as well as to recount her motives for writing a romance. In Wroth's hands, at least, romance is supple and sophisticated enough, even in its obviousness.

When she decided, most likely after the deaths of her husband and child and births of two illegitimate children, to begin the romance of the *Urania* (the first part of which was published in 1621), Wroth was not only staking her claim as a literary heir to Sir Philip Sidney but also mapping out another set of possibilities for knowing herself, literary and otherwise. Perhaps the genre's ethos and epistemology—its values that celebrate the family as a source of both powerful secret knowledge and publicly acknowledged power—supplied the best means for a "retelling" of the psychic processes that had shaped her. These values had been so crucial, after all, in establishing the first set of family ties so densely woven about Mary Wroth.[3]

Wroth was niece to Philip Sidney and Mary Herbert and the eldest daughter of their brother Robert. Married at seventeen to Sir Robert Wroth, she was a member of Queen Anne's inner circle, at least for a time, and romantically involved with her cousin William Herbert, a court favorite, notorious rake, and occasional poet. Growing up in this world meant learning from writers themselves indebted to and protective of Philip Sidney's complicated legacy. In such a world, even a decision not to write would have been a literary decision. Michael Brennan comments, for example, that many writers including Ben Jonson first regarded Philip Sidney's child Elizabeth (1585–1612) as Sidney's "most promising heir." Only after the deaths of Elizabeth and of Mary Wroth's brother William did Wroth herself become a literary figure, according to Brennan: "[a]bruptly and perhaps partly through default, Lady Mary Wroth then found herself publicly cast as a figure of central importance to an intimate group of writers, including Jonson, who regarded the preservation of the memory of Sir Philip Sidney as a means of bolstering their own literary careers."[4]

If Brennan's account minimizes Wroth's important accomplishments as a way of enlarging Sidney's, it nevertheless suggests some of the obstacles—literary and familial—Wroth faced in telling her story. Her success would involve drastically revising the story of the Sidney family, in particular, reworking lines of power and influence into patterns of filiation and likeness. I argue here that Wroth's literary efforts also involve making the chief familial tie the one between siblings rather than the one between parents and their children, so that elder sons are transformed into brothers and younger brothers become princes.

Wroth's literary ambitions are figured by the many characters whose stories circle around the loss and recovery of familial names and affections. One of them, Orilena, princess of Metelin, is betrothed to a suitor she hates, all because "he will take [her] with nothing," thereby allowing her father to bypass Orilena and make Orilena's sister his heir (*Urania* 201). The distraught Orilena describes herself as an "unblest maid," "unhappy, dispossest, and disinherited of all" (*Urania* 205).[5] Seeking help, she searches for Amphilanthus, or for Parselius, Rosindy, Perissus, Steriamus, or Selarinus, "all which," she says, "are famous men, whose honours shine equally, and either of whose assistance had been assured gaine" (*Urania* 202). In providing us with this catalog of multiple excellencies, Wroth ushers her readers into an unfamiliar romance world where heroic figures are more like than unalike and more or less interchangeable, all because they have brothers.

*   *   *

Sidney's picture of the early modern family in the *Arcadia* revealed a patriarchy hollowed from the inside out, rescued by runaway princes who restored its disinherited daughters. His narrative not only "domesticated" the conventions of Greek romance[6] but also illustrated how the genre's confusions replicate the ambiguities of families, their changing dimensions, boundaries, and loyalties. Sidney's own place in this discredited romance world was unclear, however, and he spent much of his remaining time trying to clarify his position in the unfinished New *Arcadia*, emphasizing the heroics and diminishing the "naughtiness" of his two protagonists.

Wroth takes note of the limitations of each of the worlds Sidney mapped out, but scraps the tentative solutions Sidney worked out in both cases by entirely eclipsing the world of fathers, focusing instead on the world of their sons and accounting for their lives as brothers of brothers. Philip and Robert Sidney share equal weight in such a world. But remaking the world of her family of writers is also a way for Wroth to rewrite literary tradition; such a "tendentious version"[7] of her life employs genealogy to repair a damaged romance world.[8] We see this kind of correction in "one of [her] most transparent biographical references," a variation of Wroth's own origins, as Margaret Hannay explains, in the story of her parents' courtship:

> A braue young Lord . . . second sonne to a famous Nobleman . . . was by means of a brother in Law of his, married to a greate Heyre in little Brittany, of rich possessions. This Lady was wooed and sought by many, one she affected and so much loued, as she was contented to thinke him worthy to be her husband, and so for worth, he was. (*Urania* 424)[9]

Romance magic can now be supported by history because Wroth views the family as a resource for rather than limit to desire.

Another generic adjustment is evident in the imperfect figure of Amphilanthus, charming but mercurial in his affections—"the brave Italian," he is at one point called (*Urania* 66)—whose defects and attractions suggest both Wroth's uncle Sidney and her cousin William Herbert. Amphilanthus is universally beloved, a great warrior and friend,[10] but he is hapless and unimposing, preoccupied in the narrative with the tedious business of disentangling himself from female admirers. Ultimately Amphilanthus's tremendous charms are equaled by his awkward positioning and repositioning in the story. Wroth simultaneously celebrates and deflates Amphilanthus by multiplying brothers in the *Urania*, for they are figures with imaginative claims almost equal to his own: sometimes, of course, they are dispossessed or deprived of wealth, but they are nonetheless endowed with nearly identical abilities, lineages, and appeal. These brothers are not rivals to Amphilanthus but "worthy" doubles, and Philarchos, Rosindy, Ollorandus, and others have their own stories, desires, and roles to play in the *Urania*.

Other sons and younger brothers are at once marginal and central to Wroth's narrative, perfectly embodying the *Urania*'s peculiar autobiographical rhetoric of familiarity and strangeness. Echoing Hamlet's uneasy description of his uncle Claudius, these multiplying brothers are "some thing more exactly related" to each other "then a fixion" (*Urania* 505). But representing their relations also reforms romance's "informing values,"[11] for Wroth makes equivocal what had long been taken as a given, the family's enabling institutions of lineage and primogeniture. Her challenge to what had become the "pervasive politics of the family" in the seventeenth century has manifold consequences both for early modern families and for the stories they told.[12] By focusing on the hitherto unsung stories of often slighted younger brothers, early novelistic narratives would become both domestic and elastic, more responsive to contingency and chance or accidents of birth or fate, more "realistic" and "democratic" and thus more isolated from the aristocratic politics that had idealized families by splitting them apart. Like the early novel, the *Urania* exposes the family romance as a fiction with a clear history, an unusable past, and a changing shape.

Wroth's extended universe of relatedness is the world of prose I explored in chapter 3, where connections extend in every direction in the present, and where a close-knit circle of family members is linked by shared histories and erotic partners, joint rewards, and frustrations. The same "procedures of extension" employed throughout Philip Sidney's prose oeuvre, its characteristic techniques for delineating history, documenting patrimony, keeping secrets and allocating affection, illuminate Wroth's romance world, where connections are made rather than merely duplicated.[13]

At the same time, though, Wroth's prose is burdened by lapses of family and sentiment, gaps in stories of identity and relation that appear whenever brothers come into view. Indeed, male achievement is something scandalized by family history and heroism in the *Urania* often at odds with the family's values. By definition, the family is nearly always a site of conflict and scarce resources; as Pierre Bourdieu maintains, "the closest genealogical relationship is also the point of greatest tension, and only incessant work can maintain the community of interests."[14] Bourdieu's description of the intense competition particularly sparked by brotherhood resembles the universe of the Jacobean court described by Franco Moretti, an "exemplary site of an unrestrained conflict of private interests."[15] Because the genealogical relationship between brothers "is never strong enough on its own to provide a complete determination of the relationship between individuals which it unites,"

Bourdieu claims that redress is possible "only when it goes with the shared interests, produced by the common possession of a material and symbolic patrimony, which entails collective vulnerability as well as collective property."[16] The *Urania* is the site for such redress, providing stories which link men to the family and to each other, and an altered romance world that rewards rather than challenges their ties.

* * *

The novel has been described by critics like Nancy Armstrong as inventing a narrative and a rationale for the space and values of the home, "a female stronghold" of sorts. But Wroth's unconventional romance efforts—and her "groping," perhaps, toward realism—seem to counter this trend if there is one, representing the world of the home as anything but a source of stable values, solace, or even an assured audience.[17] Pamphilia's poetry, for instance, is both encouraged and threatened by those who share her privacy, by figures like her rival Antissia (*Urania* 114) and Amphilanthus, Pamphilia's constant subject.

But part of Wroth's skepticism of family values is intrinsic to the romance genre. If romances detail the "process by which female subjectivity is brought into being within the patriarchal family,"[18] they also modify the family's "little society,"[19] sometimes by deposing its ruler, sometimes more subtly by restoring the "playspace"[20] and thereby blurring the ordinances and limits of patriarchy. The *lais* of Marie de France reveal at once the gentility and brutality of romance machinations, redefining families through forgiveness or by adultery so as to reincorporate new members and expel old ones. When romances track lost fortunes, restore missing names, or redeem corrupted identities, they also heal the breaches fathers cause. Shakespeare's Perdita escapes the death mandated by her father and king to return to him later as daughter, princess, and beloved, not only a youthful version of her moribund mother but an antidote to the narrow interpretations the claustrophobic settings of romance encourage. Rather than the plight of women in the cramped world of romance, Wroth's subject is the confines of the family.

In many ways Wroth appears to borrow from Sidney's romance in the *Arcadia* by inverting its most basic premises, detailing the experiences of female characters more or less silenced there, giving a voice and lineage to Sidney's absent shepherdess Urania, and providing a haven for the female desire and ambition rendered so terrifying in Sidney's Gynecia. As Jeff Masten puts it, "Wroth's romance begins by turning [Sidney's] world inside out."[21] Yet Wroth also offers an array of brothers for nearly all of the male and female characters in the *Urania*, siblings who shadow and support each other. "[M]ore like runne-awaies, then Princes" (*Urania* 40), these siblings foster a widened spectrum of romance possibilities and misprisions with their own stories to explain and correct romance circumstances. Wroth's introduction of these figures provides a means, I think, to explore her own unusual story as an outcast and an insider, "a Syd-ney," as Jonson put it, "though un-named."[22] Narratives of siblings supply fictions of familiar fictions, allegories of the alienation or displacement that characterizes and indeed structures families. Ultimately, the narrative world other sons and younger brothers inhabit will be colonized by the early novel. There, in the novel, private men are kept in and out of view, alternately illuminated and obscured by commonplace details rendered in a prose language of subjugation, contingency, and equivocation.[23]

* * *

On the surface of many romance texts, narrative bonds between brothers can be thin or undemanding. Malory's knightly brothers appear to duplicate each other; Chaucer's Palamon and Arcite come to blows only because Emelye has no sister. Typically the younger brothers of romance are occupied in faraway lands with ancillary adventures and, unlike the eldest brother, burdened with the requirement to live up to his example rather than to justify their paternity. When we do learn of them, younger brothers' shadowy existences exemplify Astrophil's manipulative disclaimer, "I am not I, pitie the tale of me," their adventures "mirroring fictions" that both reflect and distort reality and render patriarchy incoherent.[24] For Wroth, brothers comprise separate but equal narrative universes, and their exploits disrupt the household, revising its affective economy and political order. Later on, confusions over where families begin and end will be portrayed by the novel in stories of servants ensconced in the homes of abusive masters, but in Wroth's hands the family itself is presented as a perilous medium where identity is spread thin and lineage dangerously attenuated.

Over and over in the First Part of the *Urania* we find stories of younger brothers and other sons made to undergo exactly the same feats first brothers faced. One success story is obvious in the tale of Wroth's parents' courtship, where a second son weds a much admired lady (something at which Philip Sidney failed). Sibling accomplishment frequently serves to rectify romance's conventional mistakes, and the published *Urania* in particular portrays an entirely horizontal universe of brotherly success, failure, betrayal, and cooperation. Wroth's romance unfolds not in a timeless world but in one supported by new principles of equivalence and new emblems of contingency[25]—a world Wroth's uncle would have abhorred, no doubt, where nobility is interrupted, again and again.

For example, younger brothers frequently lose their names at the very initiation of their quests: Rosindy's lady demands this of him; Philarchos is traveling in hostile territory; or their conquests require elaborate disguises, as in the cases of Ollorandus and Leonius. Often too these brothers woo and win the hands of eldest sisters who are themselves at odds with their families. Younger brothers in this way illustrate the force of the family, the limitations of fatherly love, and the pressures of history: if the elder brother is a locus of strength, power, virility, a figure saved from the ravages of history, then the younger brother is located at the margins, a source of innovation and irritation as well as the center of an alternate telos.

In an oft-cited episode of the *Urania*, the newly recovered Princess Urania is cast from the Rocke of St. Maura into the sea by Amphilanthus (190). Urania's brother performs this "strange adventure" with much regret but with more resolve, instructed by Melissea that this is the only way to cure Urania of her love melancholy and "make her live contentedly" (190). As Amphilanthus tells Urania:

> "My dearest Sister, and the one halfe of my life, Fortune . . . hath ordain'd, a strange adventure for us, and the more cruell is it, since not to be avoyded, nor to be executed but by my hands, who best love you; yet blame me not, since I have assured hope of good successe, yet apparent death in the action, I must . . . throw thee into the Sea; pardon me, Heaven appoints it so."
>
> "My deerest brother," sayd she, "what neede you make this scruple? You wrong me much to thinke that I feare death, being your sister, or cheerish life, if not to joy my parents; fulfill your command, and be assured it is doubly welcome, comming to free me from much sorrow, and more, since given mee by your hands: those hands that best I love, and you to give it me, for whose deare sake, I onely lov'd to live, and now as much delight and wish to die." (230)[26]

Both brother and sister perform their sad parts, Amphilanthus "gently let[ting Urania] slide," yet such virtuous compliance is immediately made irrelevant by the appearance of their cousin and Pamphilia's brother, the lovesick Parselius, who jumps in after Urania. Submerged in the sea, Urania and Parselius are cured of their love for each other and promptly rescued, but Parselius has also quietly overturned the family's fatal agenda, allowing Amphilanthus to carry out the hard work of a hero while gloriously embodying what such labors typically destroy.

The introduction of the brother "remotivates" the romance structure, in Fredric Jameson's words, providing the genre with a "reality principle" at odds with its magical ones and—like Parselius's actions—supplying "a happy meanes to aide an els destroi'ed hope of rising" (*Urania* 21).[27] Historian Joan Thirsk comments similarly on the ethical adjustments younger brothers compel: "The contrast was too sharp between the life of an elder son, whose fortune was made for him by his father . . . and that of the younger son, who faced a life of hard and continuous effort, starting from almost nothing."[28] Indeed, the younger brother was the most abject of all figures in the landscape of early modern England, someone who perfectly defined the shifting affective outlines of the family and revealed its economic and political instability, for if the family's loyalties were frequently divided between father and king, its sentiments were pulled apart by home and state, duty and love.[29]

The lowly, lonely figure of the younger son is both completely produced by and wholly free of the contradictions of early modern family life, as we see in Wroth's picture of Dolorindus, a disinherited prince who demands of Steriamus the right to say misfortune is "as much mine as yours" (180). "[H]is selfe unknowne to any," Dolorindus is "an unexperienced huntsman," "having been bred abroad to learning, and to armes" (183). Despite such privations of schooling and society, Dolorindus admits to a position of rare freedom and privacy: "My selfe (though sonne unto the king, yet my sister being to inherit the king-dome) was not so much lookt after (if not by noble minds) as shee who was to rule; so as I gain'd by that meanes, both more freedome, and less over-seers of my actions" (184). Dolorindus later promises the wrathful Antissia that he will kill Amphilanthus, but then repents when he encounters his double:

> "O Amphilanthus," cri'd he, "why doe you thus exceed all possibilitie for man, how noble soever, to be a shadow to you, much lesse to equall you? Will you gather together all perfections in you to be admired, and envied by men?"

Amphilanthus is quick to forgive the prostrate Dolorindus, and then has Dolorindus marry Antissia, suggesting that Dolorindus's fortunes are his precisely because they might as well belong to someone else (*Urania* 395–97).

*   *   *

Wroth's reworking of the romance genre is not unusual in the sense that romance *always* borrows rags, disguising its aims, for instance, so that it can serve rival interests like those of the lovesick Parselius and dutiful Amphilanthus. As Northrop Frye puts it, romance is really "kidnapped" by those who employ it, stolen from its humble origins and then ensnared by cross-purposes or mixed motives, at once concerned with the home and the road away, with marriage and adultery, sin and sanctity, and all of the paths that lead individuals to and away from each other.[30] Sidney's purpose in the *Arcadia* was to make this generic waywardness a solution to romance confusions, the "truancy" or "naughtiness" of adolescent heroes he describes elsewhere in his work now providing an opportunity for those heroes to mature. At the same time, Sidney's romance exposed the family as a site of bullying and deception, tyranny and ignorance.[31] Pushing these truths even harder into aristocratic secrets, Sidney's knights in the New *Arcadia*, as Clare R. Kinney claims, translate "the lonely trials of the prototypical hero of romance into the self-serving and emphatically public endeavors of the aspiring courtier."[32] Even in the Old *Arcadia*, Musidorus's retirement from heroic labor and his disguise as a shepherd permit him liberties with a princess, and he seduces Mopsa by turning her voracious romance appetites against her, making harsh use of the stories she has repeatedly told herself.

Gordon Teskey comments on Sidney's "generous opportunism," finding "the basis of his classical values in the world of romance."[33] But the opportunism is occasionally self-interested, too, just as Sidney's heroes discover readymade loopholes and excuses in Arcadia. Wroth, while chronicling the woes of her royal lovers, challenges the aristocratic commonplaces that support many of its corroded habits of thought and desire. She reminds readers that younger sons were the ones who quested for wealth and fame and who served the heiresses never officially promised to them: romance heroes were prodigal sons who adventured precisely because they lacked homes, wealth, and a future.[34]

Northrop Frye once observed, "it was the popular Deloney, not the courtly and aristocratic Sidney, who showed what the major future forms of prose fiction were going to be like."[35] Frye's assessment fails however to acknowledge Sidney's ambivalence about the court and the aristocracy or the way Sidney's prose imagined the romance family as intimately dependent on (and threatened by) both. More than that, Sidney "defamiliarized" the family by making its members erotic rivals and its outsiders surrogate parents and forbidden love objects. Yet Sidney's story exposes the rules that govern families without finally questioning them. In contrast, Wroth makes clear the often desperate motives for romance in everyday life by turning to the prosaic, ungainly stories of younger sons. In the process, she challenges "a seemingly irresistible patriarchal family and social formation," not so much by "counterbalancing" the dominance of the "masculinist chivalric world" but by locating its cracks. The *Urania* scandalizes such a world's achievements, dismantles its logic, and reveals its origins in privilege.[36] Yet if Wroth in portraying brothers interrupts and redirects "patterns of representation," she also explores these patterns' origins in narrower underground circles, ones we typically describe—by default, I think—as domestic.[37]

The early modern novel is similarly defined by "narrative vertigo," where the main thread of a story is suddenly revealed as a digression: indeed, the position of younger brothers is repeatedly featured in works by Gascoigne and Deloney, their status one of "extraordinary instability."[38] Michael McKeon argues that their narratives should be linked with "progressive" attacks on aristocratic honor.[39] R. Howard Bloch likewise explains how romance's standard assumptions and most basic principles surrounding family life are examined in the novel:

> In the novel more than anywhere else the issues that concern the literary anthropologist—marriage, succession, narrative continuity, representational integrity, the connection between economic and linguistic property, sexual desire—are both thematized and productive of form.[40]

Wroth's long romance of Philarchos, youngest son to the king of Morea and brother to Parselius and Rosindy, is characterized by the same "narrative vertigo." She shifts the moral values and social dimensions of the romance universe by taking up his story, illustrating what patriarchy deforms in striving to guarantee its posterity, what it fails to know about itself in seeking to preserve its identity. Philarchos "pass'd through his fathers Countries unknowne" (202) searching "for a brother of his" and therefore lacking a "perfect knowne way" (201). Nevertheless, before the unknown knight champions Orilena, he encounters the deranged Princess Nereana, whom he promptly insults. Nereana immediately associates Philarchos's poor manners with low birth. " 'I am not base,' " he replies, " 'nor can I thinke you are a Princesse, since so unprincely termes come from you' " (199). Later, Philarchos will encounter Amphilanthus and challenge his romance credentials in nearly the same way. The two knights find themselves together amidst the carnage of a battlefield, and what begins as a routine exchange of chivalric pleasantries quickly dissolves into a debate about likenesses:

> "The better and the noble side," said [Philarchos], "will hardly want your company, or mine, unless wee were able to equall their valours, which I make doubt of."
>
> "It were modestly spoken of your selfe," said Amphilanthus, "but if I bee not mistaken, you might have had better manners, then to compare those together which you know not."
>
> "Why, what chollerick Knight are you," said [Philarchos], "that takes this exception, Parselius, Rosindy, Steriamus, Selarinus, Leandrus and Dolorindus, besides the brave King of Romania being there."

Once Philarchos identifies himself, however, Amphilanthus almost immediately apologizes: "I see . . . that your extreme affection rather mooved your care and haste, then ill nature." The cousins then salute each other, as is fitting, Wroth writes, for "such who were so like, as they were" (348). In relaying the younger son Philarchos's story, the *Urania* manufactures a "second edition" of the *Arcadia*, Wroth's Musidorus chastising Pamela and insulting Pyrocles; but she also produces something like a "second edition" of the Sidney family, revising its commands and prohibitions and redrawing its lines of relatedness, making her own connections to her uncle abundantly clear and completely unnecessary.[41]

*   *   *

The shadow cast by younger brothers shaped the literary landscape as well as the world outside, Joan Thirsk argues. The story of Shakespeare's Edmund, a second son who is an illegitimate one as well, makes clear the faultlines of patriarchy that all younger brothers inevitably tripped on, the gaping holes in the family's transmission of property, and commitment to posterity. Their stories repeatedly intimate that the state would falter unless the family was rebuilt affectively, politically, morally.[42] Perhaps power was increasingly exerted by the king and fathers newly in concert, so that "absolutist claims were domesticated" as Jonathan Goldberg writes.[43] But if society was regarded as an "extended family," then kings and fathers would have to find a way to better reward all their sons. Storytelling would be transformed as well. In the pictures of younger brothers who would always remain younger brothers, McKeon explains,

> the gradual discrediting of aristocratic honor, the resolution of its tacit unity into the problematic relation of rank and virtue, birth and worth, was accompanied by the accelerated mobilization of social, intellectual,

legal and institutional fictions whose increasingly ostentatious use signaled their incapacity to serve the ideological ends for which they were designed.[44]

Indeed, the development of the novel should be linked with the rising fortunes of younger sons who had been previously "sacrificed" by primogeniture "for the sake of their elder brothers."[45] McKeon details numerous "parables" of younger sons that provided an impetus for the early novel's politics and structure.[46] Both in canonical forms exemplified by the work of Defoe and Fielding and in those forms that preceded them—the novels of amorous intrigue, scandalous chronicles, and "secret histories"—stories of such figures predominate. Even Swift represents himself as "a Younger Son of younger Sons" in the midst of complaints that "the worst Part of the Soldiery [was] made up of Pages, younger Brothers of obscure Families, and others of desperate Fortunes."[47]

Dolorindus, Philarchos, and other younger sons share many motives with early novelistic prose's band of pirates, criminals, pilgrims, and other prodigals.[48] Bersindor marries above his station (*Urania* 499) like the Courtly Forrester so envied by Amphilanthus, who marries a Nimph (345). The prince of Istria (the king of Dalmatia's brother) steals Urania as a baby (231–33). Leonius, Amphilanthus's brother, disguises himself as the forest nymph Leonia (432–36). The black knight Rosindy disguises himself as a servant to wait on the beautiful Meriana and thereby "[imagines him]self more then a Prince, in being so happie to be his servant to such an end" (108). Rosindy is elsewhere taken for a "supposed Amphilanthus" by Antissia (115–16).

Other younger sons have talents older sons lack: Seleucius, brother to the king of Romania, cares for Prince Antissius after his stepmother's betrayal (*Urania* 51–59). Polarchos, bastard son of the king of Cyprus, rescues Parselius and Leandrus (405–06). Many of these figures come to Amphilanthus's aid in particular. Ollorandus, the Forrest Knight and son of the king of Bohemia, tells Amphilanthus his story, Amphilanthus pledging it "shall binde me to confidence, and ingage me to the relation of mine" (78). Polarchos likewise preserves Amphilanthus's role in the *Urania*, returning Amphilanthus's armor to him at the end of Wroth's story. In fact, the romance concludes with Amphilanthus's praise of Polarchos as "noble Polarchos," "the bravest friend that ever man had" (660).

Wroth's representation of a now-ambiguous moral universe pits conventional romance values like lineage or status against novelistic virtues of ingenuity, persistence, and character.[49] More than that, the *Urania* gives us a moral and ontological framework for understanding younger sons' behaviors and aspirations by representing all males as really more like than unlike. The generic consequences are extreme, and the *Urania* takes shape in "a strange and unfamiliar world of symbolic significance"[50] where romance heroes are merely "simple abstractions"[51] detached from the families and histories that motivate them. Wroth maps out in all its diffuseness the political, amorous, and psychological arrangements of this new world, a project that also characterizes so much of later women's fiction, the "secret" histories, lives, and adventures that novelists like Richardson, Fielding, and Defoe would seek to eclipse.[52] The family is newly made whole in Wroth's hands, unveiled as a cosmos in itself, a site of myth and of reality all at once.

* * *

"It is impossible," Maureen Quilligan notes, "to overstress the strange plethora of brothers and the even more bizarre patriarchal silence of fathers in the *Urania*."[53] Wroth's radical representations of ruling women and nurturing female friendships in the *Urania* often mitigate the picture of Pamphilia's passive constancy, but these images do not finally challenge patriarchy where it lives, in any existential or deep way.[54] More threatening are the lengthy and often inconclusive spectacles in which younger brothers play such formative parts, which stand in marked contrast to the "proliferation of patriarchs" in Shakespeare's plays.[55] As Anne Shaver observes, "[e]ach of Pamphilia's and Amphilanthus' younger brothers acquires a kingdom by marrying a ruling queen or an heiress."[56]

Yet the acquisition process is a highly subversive one. Ollorandus had been refused a knighthood by his father because, the young man tells us, "my elder Brother, being weake and sickly, had not demanded it; resolving I should attend his encrease of strength, my Fathers whole content being in that Sonne" (78). As in other episodes, this second son persuades his beloved Princess Melasinda to marry his rival Rodolindus, her uncle king's bastard son (79). The Forrest Knight explains that "[w]ith much adoe, and long perswasions I wonne (her love to mee) her yeelding for the other; so the match was concluded, and peace on all sides, I leading her the day of her marriage to her wedding Chamber, where I left her to her husband" (80). But Ollorandus's "perswasions" are also rewarded by a garden assignation: "Thus was I the blest man," the Forrest Knight reveals, "injoying the world of riches in her love, and [her husband] contented after having what he sought" (80). Even here, two disinherited sons can enjoy a union with a queen.

Doubtless her uncle's example as frustrated courtier and royal favorite made Philip Sidney a formidable model of and inspiration for Wroth's reworkings of family life.[57] Yet it is Wroth's father Robert's position, as younger brother of Philip and Mary Sidney, which supplied a more persuasive motive or challenging riddle, that of the poet without a vocation, the male figure without a role. Gary Waller claims that "Wroth's life shows a consistent struggle to come to terms with her idealization of the figure of the distant father."[58] I would only add that Wroth could explore and explain this distant father's authority by using romance to map out his equivalence to Philip Sidney and proclaim herself the daughter of a younger son. Critics who link the ambitions of romance with those of the early novel have intimated such a project to remake brotherly positioning:

> The French term *roman* of course designates a more heterogeneous group than the works that English distinguishes by the terms "novel" and "romance".... But the distinction between "romance" and "novel" is rather unstable, especially with respect to the French *romans* of the twelfth and thirteenth centuries.... Literary history, as Shklovsky said, is the history of the successive murders of fathers, or attempts at reinstating uncles.[59]

To be sure, Wroth drew on other writers besides her uncle and father for source material and narrative principles.[60] But the interest in the family as a source of romance problems and their solutions is Sidney's particular contribution to the genre, as is his abiding interest in giving family relations a place in history and a shape in literature. Still, while for Sidney the romance world of the family was the "nursery of the heroic,"[61] Wroth suggests that there are peculiar comforts and long-standing disturbances that organize the home and occasionally force some members outside.[62] Exiled in this new region without language or symbol, brothers are given a chance to correct the mistakes of fathers and take possession of the space overlooked by elder brothers, to conquer an "abyss" upon which all representation finally rests.[63]

*   *   *

Many readers have responded to the world of the *Urania* with more description than analysis of its poetics and logic. Mary Ellen Lamb explains this response in terms of the *Urania*'s "refusal to cohere": "To write about the *Urania* at all," she rightly claims, "requires a form of rewriting: a selection of events according to a reader's own critical agenda to create a coherent pattern. Of necessity, this artificial coherence distorts a central feature of the *Urania*."[64] Still, we might see the multiplication of plots and characters as part of a larger project to reorient our reading of the romance landscape, to privilege space over time, metonymy over metaphor, brothers over fathers, so that genealogy becomes a work in progress, without beginning or end.

This fluid romance patterning contrasts with the linear movement of epic narratives where there is, according to R. Howard Bloch, a clear-cut relationship between "genealogical succession and narrative structure," between power and knowledge. Even the "*chanson de geste*," Bloch claims, represents "an act of symbol production affirming the linear organization of the feudal family as deeds, swords, and . . . word meanings pass down lineal lines."[65] Bloch adds: "courtly romance constituted a privileged site for the mediation of conflicting grammatical, familial, and literary models" so that history would coincide with familial identity and epic rivalry with brotherhood.[66]

Amphilanthus's heroic failures reveal the flaws in this romance project, embodying the genre's basic principles of scarcity and obsolescence. But the younger brother provides a solution to this romance dilemma: he is a mechanism for mediating between less pliant *semes*, discrete pieces of meaning like elder brothers, beautiful princesses, and distant fathers impervious to romance designs. The younger brother intervenes in this world, enabling the family to extend rather than implode, so that romance can conclude with regeneration.[67] The Princess Urania explains Wroth's alteration of romance patternings as an affective one, recasting the psyche and desire and their relation to the world outside. Urania is explaining how her love for Parselius was replaced by an equal love for Steriamus:

> it was mee thought a wonderfull odde change, and passing different affection I did feele, when I did alter: for though I were freed from my first love, and had a power to choose again, yet I was not so amply cured from memorie, but that I did resemble one newly come out of a vision, distracted, scarce able to tell, whether it were a fixion, or the truth; yet I resolved, and so by force of heavenly providence lost the first, and live in second choice. (*Urania* 331–32)

Wroth's narrative is one of second chances and multiple choices, the limited possibilities proposed by the "ideal" and "failed" romances Janice Radway describes now replaced by alternatives that are perfect because they are equivalent: as Urania puts it, "thus if I changd, twas from sweete Steriamus to Parselius, for his excellency wonne me first; so this can bee no change, but as a booke layd by, new lookt on, is more, and with greater judgement understood" (333).

We see such a new romance economy—what Wroth calls "new worke in new kinds" (231)—articulated later in *Urania*, where the "Lady of the oddest passion" loves two knights equally (453). The threesome jointly conclude, "one woman might love two men lawfully, and constantlier then one, and that it were much safer for a man to have his friend his Rivall then to be alone" (451). Ollorandus, sitting in judgment, seems to concur: "mutuall affections were better held still as they were," he states, "lest division might divide the perfect love between these friends" (449). Reorganizing the psychic process through stories of interchangeable men, Wroth offers "the possibility of sensing other historical rhythms," another way of building families and continuing their private designs.[68]

But in Wroth's iteration of brothers we also have the dismantling of the romance, now split into what Fredric Jameson calls "syllables and broken fragments of some single immense story," where "plots of affiliation" replace "plots of power."[69] After all, Wroth has altered romance's most basic perspectives and values, inverting what Jameson calls its "presupposition," "the ethical axis of good and evil," which had itself rested on a problem of chronology. As Jameson explains:

> Romance in its original strong form may then be understood as an imaginary "solution" to this real contradiction, a symbolic answer to the perplexing question of how my enemy can be thought of as being *evil*, (that is, as other than myself and marked by some absolute difference) when what is responsible for his being so characterized is quite simply the *identity* of his own conduct with mine, the which—points of honor, challenges, tests of strength—he reflects as in a mirror image.

Previously romance "[had solved] this conceptual dilemma by producing a new kind of narrative," "the 'story' of something like a semic evaporation" whereby, Jameson argues, the evil knight typically is defeated, reveals his name, and is ultimately "reinserted" into the unity of the social class, thereby "los[ing] all his sinister unfamiliarity."[70] But in the *Urania*, brothers are not lost or evil but temporarily displaced, and their reappearance always signals that the family is a more capacious if disorganized structure, with many hidden rooms and many mirrors in them.

*   *   *

Lady Mary Wroth enjoyed a prestige and security to which Philip Sidney always aspired, "a social legitimacy even her uncle did not possess," as Maureen Quilligan observes.[71] She was the eldest daughter of an earl held in long esteem by James I, and she married into even greater wealth and status. For many years Wroth was also a member of the innermost courtly circles and a particular favorite of Queen Anne, at least until the birth of two illegitimate children after the death of Wroth's husband in 1614. Yet this social legitimacy was accompanied by the economic and psychological dislocation all early modern families experienced, a chance to renew themselves around new principles of affection or connection or find themselves entirely dissolved and incoherent.

Wroth's fascination with the figure of the younger brother is suggestive in its outline of the challenges faced by her father, a figure for many years utterly removed from the politics and social rules so crucial to his fortunes and erotic life. His—and not Philip's—figure perfectly defines the early modern family in its crudest dimensions and most essential components. To be sure, there are many contradictions of affection that enable some children at the expense of others, and the stories of Cain and Abel and Joseph remind us that the Bible is literally riddled with these problems. Such images of failed or lost authority and of families condemned to "un-knowing" each other particularly motivate the early novel. But Wroth's repeated images of brothers also underscore a belief that the family is *not* a public unit nor an essential link between persons;[72] the power of the family instead is "always piecemeal and discontinuous," always "operating within the interstices of a culture to provide the illusion of overmastering structures."[73] In the novel indeed, masterful men take their places alongside masterless men, and firstborn sons are regularly pictured right next to their younger brothers.

This picture of an extended and decentered universe is a hallmark of later women's fiction, which is typically described as diffuse, baroque, or distended. The connections to Wroth seem especially clear in Aphra Behn's later fiction, where the term "brother" is a euphemism for "fiance" or "pimp," and familial relations a cover for shattering betrayals of trust and feeling. If the family has at this point become a concrete world of "definite and limited achievements,"[74] "scandal [has become] the defining feature of affective relations."[75] Christopher Flint further explains: "Domesticity is the supplement to other social mechanisms of power rather than the central repository of cultural values."[76] The heroines in Behn's "The Unfortunate Happy Lady" (1684) and "The Wandering Beauty" (1698) display their virtue, for instance, by rejecting the immediate family at the outset of the story or reworking it in accordance with their own wishes at the story's end, and Behn specifically attacks the institution of primogeniture in her comedy unambiguously titled "The Younger Brother; or, the Amourous Jilt" (1684).[77]

Nonetheless, it is the plasticity of familial relations that allows families and romance, too, to endure. After 1600 the romance, according to Michael McKeon, became a form notable for its "incapacity" "to serve any relevant ideological ends."[78] Perhaps this is because family life became increasingly restricted to the domestic sphere, itself a site of invisible violence, privately monitored values, and unremarkable affections, along with passions no less fierce or formative in the secrecy that surrounded and shaped them. But perhaps romance was simply driven underground once more; as Marthe Roberts explains, in the psychic sphere of the family are "all the ingredients of an unformulated novel, of an incipient fiction."[79] At the same time, the family secrets of the Sidneys illuminate a larger private world that would increasingly find itself in the novel. The dead Philip denied Elizabeth's maternal affections would become a Protestant saint, and his brother Robert, an admired courtier much honored by King James I, Elizabeth's rival's son. A generation later, William Herbert's rare influence with Queen Anne was only rivaled by his brother Philip's influence with the king: "As [the Queen] had her *Favourites* in one place, the King had his in an other. She loved the elder Brother, the Earl of *Pembroke*; he the younger, whom he made Earl of *Mountgomery*, and Knight of the *Garter*."[80] These contests between brothers, endlessly repeated, suggest the ways familiar ties are continually reworked: always metaphorical, always contrived, such links between family members can never thus be broken.

EPILOGUE

"CURIOUS NETWORKS" AND LOST SONS

What else can be said about a thirty-one-year-old poet who died? Much of the poetry written subsequent to Philip Sidney's death—what we commonly describe as the English literary tradition—wrestles with this problem. Edmund Spenser's efforts are instructive, for he sometimes construes Sidney as a father figure, at other times as lost son. In *Muiopotmos*, a short fable published in the 1591 volume of *Complaints*, Spenser transforms his dead mentor Sidney into a gorgeous insect:

> Of all the race of silver-winged Flies
> Which doo possesse the Empire of the aire,
> Betwixt the centred earth, and azure skies,
> Was none more favourable, nor more faire
> Whilst heaven did favour his felicities,
> Then *Clarion*, the oldest sonne and haire
> Of *Muscaroll*, and in his fathers sight
> Of all alive did seeme the fairest wight. (ll. 17–24)[1]

Perhaps the metaphorical connections between maimed poet and "unhappie happie Flie" (l. 234) seem strained, but no more so, Spenser implies, than the way Philip Sidney's heady felicities continually tested the constraints of the Elizabethan world.[2]

Other makers of the Sidney legend, as I have explored in this book, concentrated on Sidney's courtly talents, his effortless grace or *sprezzatura*, his piety and learning, his artistry and warrior courage. Spenser focuses instead on the short life and sad death of the author. But there are other connections. The same incredulity attends our reading Fulke Greville's account of Sidney throwing off his leg armor before the fatal battle at Zutphen and Spenser's picture of the "faultles" butterfly Clarion arming himself with breastplate and "hairie hide of some wilde beast" (ll. 57–66). The same epic incongruity clouds both of these stories and exposes the faulty assurances of swollen myths that brace them.[3]

Spenser's unusual treatment of Sidney is the exception in a long stream of unsullied tribute from Sidney's survivors, friends, and other poets in his wake—an anomaly in the elegiac tradition that supplies us with the picture of Sidney still held even today: "Spirit without spot," generous patron, national hero, devout Protestant and worldly knight—St. George and Adonis, Lycidas and Astrophel—the realization of Petrarch and Castiglione's blueprints and a model for Shakespeare's Hamlet ("th' observ'd of all observers"), Elizabeth's rival and thorn and conscience. All of this mythmaking ignores Sidney's discomfort with his court, his queen, his poetry, and himself, however.[4]

Indeed, Sidney's repeated indictments of myth's facile assumptions have little in common with posthumous efforts to make *him* a Protestant saint or perfect courtier, efforts like those of Greville, his closest friend since boyhood, from whom we still draw much of our biographical data. More dogma or hagiography than myth—since it permits no variants—Greville's glorious narrative in *The Dedication to Sir Philip Sidney* (1652; likely drafted 1609–14) provides a kind of "all-purpose ideological epoxy"[5] to seal the cracks in Sidney's broken image and repair any gaps in Elizabeth's mercurial reward system as well. Greville's magical compound is powerful enough, too, to retract Sidney's deathbed retraction of the *Arcadia*:

> When [Sidney's] body declined, and his piercing inward powers were lifted up to a purer Horizon, he then discovered, not onely the imperfection, but vanitie of these shadowes. . . . And from this ground, in that memorable temperament of his, he bequeathed no other legacie, but the fire, to this unpolished Embrio.[6]

For Greville, death mocked Sidney's body but not his aspirations: it divulged his failure to finish things ("this unpolished Embrio") rather than his ability to create. Furthermore, Sidney's "piercing inward powers" literally preserve his authorship by showing him, at the last, to be making ruthless editorial decisions. In the dedicatory letter to Mary Herbert, Sidney had called the *Arcadia* "this idle work of mine" and likened it to "the spider's web," "thought fitter to be swept away than worn to any purpose."[7] Luckily, Greville, to whom Sidney had entrusted his working copy, disregarded his friend's wishes and held onto the manuscript.

* * *

Spenser's imagination is a particularly sturdy vehicle for the Sidney story, well suited to handle the discrepancies in his predecessor's image because it is roomy enough to manage tensions between polite chivalry and politic ambition, Christian virtue and courtly failure. Book 6 of *The Faerie Queene* is an especially detailed catalog of such pressures, sensitively registering a legendary knight's shortcomings and potential. Spenser's efforts in *Muiopotmos* are more experimental, however. As one critic notes, "the poet has adjusted and sometimes disoriented our view of Clarion, variously collapsing or reinforcing the distance between man and butterfly."[8] Here Sidney is more of a stumbling block and less of a hypothetical solution, but *Muiopotmos* also illustrates exactly what Spenser learned from a model so expert at mutating his own troubled identity, especially "the shape of his life *as a poet*."[9]

Sidney's integrations and disintegrations of identity take center stage as Astrophil or Philisides, although they could also be unnamed, like the courtier turned horse in the opening of *The Defence of Poesy*.[10] Such recurrent (and often casual) alterations in the poet's "shape" may have encouraged Spenser to describe Sidney as a butterfly or his avocation in terms of pastoral tragedy; yet it also took time and skill for Spenser to fashion his careful portrait of a "defamed gentle Knight" (*FQ* 6.12.34.6), one reminiscent as much of Redcrosse in his inaugural battle with Error as of Calidore in his futile war against the Blatant Beast. *Muiopotmos* presents us with the imaginative difficulties that Spenser was up against and makes the intractability of the problem a feature of his framework.

A later example of these difficulties is found in *Astrophel*, a 1595 elegy for Sidney, where Spenser's grief is muted and almost silent on the subject of Sidney's literary legacy. Focusing on Astrophel, "of gentlest race that ever shepheard bore," a figure who "could pipe and daunce, and caroll sweet" (ll. 2, 31) Spenser's poem shares with Greville's account the sense that Sidney's frustrations with his art more deeply marked a personal dissatisfaction with himself as an "unmynd-full," "lucklesse," "wretched" "boy," "not so happie as the rest" (*Astrophel* ll. 112, 142, 133, 12). In Spenser's retelling of Sidney's Petrarchan woes, Astrophel courts Stella not "with ydle words alone" or "verses vaine (yet verse are not vaine)"—as Spenser silently corrects the presumptuous lover—"But with brave deeds to her sole service rowed" (ll. 67–69).

Yet in Spenser's view Astrophel's folly is compounded by his bravery: he is "too hardie alas" (l. 72). "Unlike other pastoral elegies," Theodore Steinberg observes, "*Astrophel* does not even pretend to offer consolation. Instead it both mourns and blames the deceased."[11] Many readers in fact protest *Astrophel's* "alleged lack of passion," but Spenser stands apart from contemporaries who were "seduced," Steinberg claims, by the Sidney myth.[12] Yet Spenser's *Astrophel* does not admonish or trivialize its subject as subtly or as forcefully as the earlier *Muiopotmos* does. Only *Muiopotmos* seems subversive enough to do his failed model's memory justice.

Still, Spenser's poem about a butterfly does not reject myth altogether in its picture of Sidney. Instead, Spenser shapes a myth about what Paul Tillich calls a "broken myth," constructing a narrative obviously "understood as a myth" but "broken in its immediacy or literalness," with its "inherent symbolic, evocative, revelatory power [nonetheless] intact."[13] Clarion's carcass finally severed from "the spectacle of care" (l. 440) seems an obvious example of a "broken myth": a fly turned fragment, a hero nourishing spiders. But Spenser's story also diminishes the figures of Jove, Achilles, and Icarus along the way. They are evoked only to suggest beasts like bulls and flies (see ll. 277–79, 63–64, 41–48), their heroic instincts and motives foiled by an equally powerful—and equally banal—set of directives. Spenser's narrative framework and imagery and allusions in *Muiopotmos* all ironically confirm the ordinariness and routinization of a story of flight and fall, of ambition and consumption, in an epic of the poet's short rise and necessary eclipse.

* * *

Because Sidney and the butterfly are indices of Renaissance poetry's mythical capacities, both allow Spenser to question the meaning of his own literary legacy, "whether it amounts to anything more than the sum of its parts, and whether it will continue to exist."[14] They even serve as a way to test whether Spenser's ideas of himself are as mythical as those fond conceits surrounding Sidney: to determine, in other words, whether poets are parts of the world or fragments of its gratified imagination. We see this testing inaugurated at the outset of *Muiopotmos*, where Spenser breaks apart the Sidney myth almost immediately. Even a reader's very first impression of *Muiopotmos* is a wrong one.[15] The narrator promises to sing of "deadly dolorous debate" "[b]etwixt two mightie ones of great estate" (ll. 1, 3) but the combatants are not human; moreover, the insect-hero Clarion is blissfully unaware of the hostility of nature or that his own nature is part of a larger, longer history. He is oblivious to the limits of a world that only appears designed for him. Waldo McNeir characterizes a similarly deceptive Tudor world in his reminder that Queen Elizabeth

> brooked no competition from the possible makers of their own personal myths . . . poor Mary Stuart, who bungled everything she put her hand

to except the embroideries on which she whiled away her captivity; Walter Raleigh, brilliant but unlucky, a puzzle to himself . . . Robert Dudley, Earl of Leicester . . . or glamorous Robert Devereux, Earl of Essex, a foolish hothead, a self-important popinjay. All these potential luminaries of Elizabeth's reign were programmed to self-destruct.[16]

The possibilities for "self-fashioning" in both worlds appear utterly illusory, because myth—even Elizabeth's myth of herself—is the kind of narrative over which a subject has little control. At the same time, however, ignorance or oblivion can have its own heroic dimension on paper and in real life. Richard Lanham comments: "Sidney's blindness to the practical compromises (and sometimes devious tricks) of those closest to him indicates his suitability for resurrecting and personifying a chivalric ideal irrelevant to the world of affairs. In his blindness lay his virtue and his real usefulness—as the ornament of his age."[17] We might go still further, and propose that Sidney is the implied reader of *The Faerie Queene*, Elizabeth only an available substitute. Sidney's anachronistic talents and pioneering gifts would have been especially well suited to evaluate the delightful, treacherous realm of faery.

Clarion's gorgeous world is similarly insecure, though every "cosmic calamity" turns out to be the product of careful design.[18] Indeed, Clarion's image is continually elaborated and subverted by the poet: he serves at once as innocuous fly and crown prince (l. 32); embodiment of Achilles and Icarus; armigerous insect and painted peacock (l. 95). This splitting is another way of reminding us that the butterfly is ultimately fictive, part of some other narrative or a remnant from another battle over mythmaking. These limits explain as well why Clarion's reality keeps getting ambushed (see l. 404).[19]

Yet Spenser also details the culpabilities of his heroes. Clarion's vulnerabilities are both internal and external: he is faultless, but careless, innocent but presumptuous. Lanham intimates we might view Sidney's vulnerabilities in the same light, pointing to Sidney's response to Elizabeth's offer of recusant fines after he had applied to her for preferment. Sidney wrote in anguish to Leicester: "I know not truly what to sai since her Majestie is pleased so to answer, for as well mai her Majestie refuse the matter of the papistes and then have I both shame and skorne."[20] Sidney could organize a world that sometimes spun mercilessly about him because his "shame and skorne" were something he might calibrate beforehand. His image provides us with a figurehead cracked in two, a reluctant begetter of a courtly style and artful logician of a moral code. If Spenser's image of the butterfly serves as a symbol of mutability and mortality, the soul and the imagination, a tiny insect and an overweening god, it also works as a rhombus or chimera—something that's been turned or warped or grafted—or what Fredric Jameson calls an *ideologeme*, a fragment of an older discourse.[21] For this reason Spenser multiplies the stories of Clarion in his poem, imprisoning the hapless butterfly again and again in his art, experimenting with alternative cosmogonies or accounts of origins.[22] Sidney's mythical status is now subject to a variety of historical pressures, so that the myths about him are part of Spenser's myth about him.

This epistemological influx is a recurrent feature of Spenser's work ethic. As William Blisset claims:

> Spenser is the poet of the second thought: his normal practice is to arrange polarized or complementary episodes and persons so as to present a certain total conception. The deepest and strongest impressions are made first in *The Faerie Queene*, but they are followed by finer ones calculated to right any imbalance of feeling.[23]

Such "second thoughts" about Sidney are as obvious in *The Faerie Queene* as they are in *Astrophel* or *Muiopotmos*, where again and again Spenser supplies an image of Sidney simultaneously splintered by and relieved of contradiction, rescued from doubt and subject to interpretation. *Muiopotmos* supplies the most complete discharge of anxiety, however, making Sidney mythical by making him inconsequential. As Robert A. Brinkley puts it:

> As gods we celebrate the artistry which produces Clarion and Aragnoll as insects; as insects we feel diminished by that artistry. Yet Spenser's poem may involve a third perspective as well, that of a reader-narrator who, by understanding the fate of the butterfly, absents himself from that fate.[24]

* * *

The scandalous collapse of lyric at Spenser's Acidale, the ambiguous cosmic solution in *The Mutabilitie Cantos*, and the abrupt rupture of texts in the library atop Alma's Castle all illustrate epistemological crashes, where ways of viewing the worlds Spenser creates fall suddenly and irrevocably apart. *Muiopotmos* is the most abrupt and graphic treatment of such a breakdown. In lieu of the "endless reduplication" Angus Fletcher describes as constantly at work in *The Faerie Queene*,[25] Spenser now fashions a discrete series of vanishing points and terminal positions, places where perspectives interrupt and subvert or even consume each other. One reader has suggested that *The Faerie Queene*'s heroes have much in common with Clarion, since they "[tend] to be a reproduction of heroes from Spenser's sources: Arthegall dressed in Achilles' armor is a serious version of Clarion clothed in Achilles' breastplate." Such reproductions emerge out of "'an absolutely recreative style' which dislocates Spenser's sources by placing them in contexts which distance the displaced material from the meanings they represented."[26] But *Muiopotmos* repeatedly pushes myth against a wall, emptying out mixed messages and forcing endings out of beginnings. The result is that Clarion's mythical genealogy simultaneously explains and condemns him.[27]

What else can you say about a failed poet? Spenser's career was marked by sustained meditation on the virtues and projects and later the sorry demise of Sidney, the "president / Of noblesse and of chevalree," as the New Poet puts it in *The Shepheardes Calender* (1579) (ll. 3–4). One may take issue with Ernest de Selincourt's assessment that "Spenser's love for Sidney was probably the deepest formative influence upon his life and character,"[28] but Spenser's professional and personal debts were continually felt, if ceaselessly interrogated.[29] Yet while his ideas about Sidney as "perfecte paterne" and "special favourer and maintainer of all kind of learning" seem relatively uncomplicated in *The Shepheardes Calender,* Spenser's sentiments and questions grow increasingly ambivalent, as one can detect in the tortured form of the Redcrosse Knight, the Protestant hero of *The Faerie Queene* (1590). Both "clownish younge man" and embryonic saint, Redcrosse is "carelesse of his health, and of his fame" (1.7.7.3).[30] He is neither a confident arbiter of taste nor reliable judge of values, but a frail warrior tossed about by dreams and doubts, a flawed and occasionally hysterical hero more sinning than sinned against.[31]

The Faerie Queene provides steady exercise in recognition and reappraisal, in what Blisset calls "second thoughts," educating readers and characters alike in the intricacies of fictions, myths, and mistakes as variously embodied by figures like Una and Duessa, the two Florimells, and the disguised Archimago. False dreams and faulty texts offer additional practice. Redcrosse Knight's victories occur when he can credit his second thoughts about Una, Duessa, Despair, and the apocalyptic dragon, and such generous "revisionary play" forms the substructure of the mythology Spenser weaves for Sidney, too, where, as Harry Berger explains: "Every triumph or resolution at a lower level of existence or an earlier phase of experience releases new and different problems at a higher level or a later phase."[32]

This experimentation becomes even clearer in *The Faerie Queene*'s second installment (1596), where we can make out Sidney's new form in the politely condescending figure of the Knight of Courtesy. Calidore's slumming in faeryland could be an image of Sidney retired at Wilton (or banned from court): held captive in the poem by his love for Pastorella, Calidore completely relinquishes his epic duties, forswearing the court and its "shadowes vaine" for a "another quest, another game" (6.10.2.3). This smugly obsequious Calidore reveals the sheer frailty of poetic aspiration when he tramples on it. Greedy and "Vnmyndfull" (6.10.1.3), Calidore ruins Colin's celestial circle and neglects Pastorella for yet "another game in vew" when he spies on the maidens dancing in a ring to Colin's pipe. In the guise of the Blatant Beast, Spenser exposes myth as an open secret, the dance of the graces at Acidale merely a broken circle about a "poore handmayd," with "one minime" taking the place of Gloriana's deserved song (6.10.28.6).

Spenser takes these poetic failures even further in *Muiopotmos* since heroic action is altogether impossible in its universe: the actors are insects, their struggles instinctive, their origins arbitrary, their outcomes completely natural. His qualifications of heroic myth and familial bonds are emphasized by two framing stories about mothers, both of which reveal mythical origins and precedents as anything but self-evident or inevitable. Spenser provides three drastically different accounts of myth-making in *Muiopotmos*, offering a comic tale of a spider and butterfly along with accounts of their respective origins.

The two framing narratives focus upon mothers and divine anger, and make genealogy the product of prejudice, history the result of art, and nature the end result of mythmaking, something "golden" and "brazen" all at once. These poetics of maternity thus provide no dynastic guarantees; if anything, they subject sons to the fiercer pressures of biology and raw power—the "absolutism of reality" that Hans Blumenberg describes. Spenser's first story is a tale of hermeneutical suspicion or misprision in the conflict between Venus and Astery. His second features a story of artistic rivalry and one-upmanship in the account of Arachne and Athena. The third tale, which develops out of those antagonisms, presents a story of readerly ignorance and writerly craft in the ensuing contest between Clarion and Aragnoll. It is through this mythical inter-section of motives and fears that Spenser can demythologize Sidney, exposing how history, nature, and storytelling are all equally fabled, and lineage—whether poetic or ancestral—is subject alike to corruption and expediency and, in Sidney's case, more or less a dead end.

Spenser's tale of Astery is his own invention. We learn that Venus is still reeling with jealousy over the affair between her son Cupid and Psyche and susceptible to rumor when Astery surpasses the other nymphs in collecting flowers for the goddess. This first narrative seems a perfect example of what Wallace Stevens calls a "local myth," whereby casual inflections and innuendoes carry tremendous causal power. Believing her nymphs' malicious gossip, Venus angrily turns Astery into a butterfly, "And all those flowres, with which so plenteouslie / Her lap she filled had, that bred her spight, / [Venus] placed in her wings, for memorie / Of her pretended crime, though crime none were" (ll. 140–43). Clarion's genetic history stems from this maternal furor: his origins are at one with celestial reaction-formation, his beauty and fragility a knee-jerk response to a mother's worries about her own reproductive powers.[33]

But this dynastic upset also resembles the story of Aragnoll's origins. Aragnoll's mother Arachne, the "most fine-fingred workwoman on ground" (l. 260), rashly challenges Pallas Athena to a weaving contest "to compare with her in curious skill / Of workes with loome, with needle, and with quill" (ll. 270–72). Arachne creates a splendid tapestry of Jove raping Europa, "full fit for Kingly bowres" (l. 300), artfully transforming mythological violation into a royal wedding gift and reconstructing mythical violence as the reproduction of history. This is a canny way for Arachne to reduce her rival's divine power by placing it within a context of brutal manipulation, miscegenation, and trickery. She not only naturalizes the divine, but demythologizes her opponent's history. Indeed, Nancy K. Miller reads the Ovidian episode as centering upon "the spider artist" whose story "thematizes explicitly the conditions of text production."[34]

But Renaissance mythographers had different ideas about such conditions, and we might also remember what Sidney had to say about needlework in *A Defence of Poesy*:

> Truly, a needle cannot do much hurt, and as truly (with leave of ladies be it spoken) it cannot do much good: with a sword thou mayst kill thy father, and with a sword thou must defend thy prince and country. (55)

Sidney links maternal preoccupations with inactivity or inefficacy, and male bravery to patricide; in either case, the actor's accomplishments are self-defeating. In Spenser's hands, too, a revolutionary tale is reduced to a story of feminine failure and arrogant effort, since Arachne's beautiful transgression pales next to Athena's storytelling. Athena not only weaves her own divine history into myth, "the storie of the olde debate, which she with *Neptune* did for *Athens* trie" (ll. 305–06), but trumps it in Spenser's revision: "[w]ith excellent device and wondrous slight," she places a "[b]utterflie emongst those leaues" "[t]hat seem'd to liue, so like it was in sight" (ll. 330, 329, 332). Athena makes the banal heavenly and recovers a nature buried in the "leaues" of fiction while subtly reminding her audience that insects roam royal corridors and flies swarm empires. She further indicates that mythmaking persuasively reproduces history through means other than violence.

Both framing stories also present us with the machinery of divine authority as something engaged whenever the gods generate supplements or local myths, fictional fragments that hold other older stories in place. These new stories rework and suspend reason, altering ontology by reminding humans of it, introducing new "limit situations" by addressing old ones. The "astonied" and "dismaid" Arachne, for instance, is filled with envy over Athena's tapestry and disembodied by bitterness; "her blood to poysonous rancor turne," she becomes a spider (ll. 339, 341, 344). In both stories of warped origins, the mothers' descendants will live out the painful legacy of local myth, where one imaginative universe stops suddenly short of a second. And, for the second time in the poem, the butterfly is revealed to be nothing but a by-product, something "overlaid, / And mastered with workmanship so rare" (ll. 337–38).

In the final story, Aragnoll, mindful of his mother's grudge, weaves "straight a net with a manie a folde / About the cave . . . / With fine small cords about it stretched wide, / So finely sponne, that scarce they could be spide" (ll. 357, 359–60). Aragnoll is a neat precursor for Milton's serpent, the "foe of faire things, th' author of confusion" (l. 244); and here as in *Paradise Lost* art is at its pinnacle when it becomes invisible, an unpredictable or unseen force of history. It is far beyond the ken of the "careles" (l. 375) Clarion to acknowledge, much less avoid. The butterfly is at once destroyed by and preserved in Aragnoll's web, a "curious networke" (l. 368) of envy, resentment, artistry, and instinct, a space where golden and brazen worlds magically collide.

* * *

Sidney's brief career was continually disturbed by anxiety over how he would be perceived by readers and nonreaders alike, his poetry frequently supplying a coy set of inflections and innuendoes to offset the rude noise of rumor, slander, and courteous disdain. Wallace Stevens found in such a harsh world a music of its own, the thin music that comes from "the blackbird whistling / Or just after." But Sidney never experienced either song, never having found a way to bridge the gaps in his experience or the holes in others' pictures of him: for Sidney there was no single style or perspective or set of beliefs that would provide either stability or escape.[35] Spenser makes this particular failure clear in the hapless butterfly's death at the hands of Aragnoll as well as in the sad transaction at Acidale, when the piping Colin is interrupted by the knightly Calidore. Most critics see in Colin a figure for Spenser and in Calidore one for Sidney, but the two figures confront each other in Spenser's poem, each exposing the other as unnatural or irrelevant, something superfluous in their respective worlds.[36] Left with a broken bagpipe, Colin seems both a relic and a new model for the New Poet. "Not I so happy," he tells Calidore, "As thou Vnhappy" (6.10.20. 1–2).

Milton will ransack this murky legacy in his pastoral elegy "Lycidas" (1638), turning his "gentle swain," fallen poet, and favorite son into Orpheus, the Muse's "inchanting son" (ll. 58–59), "sunk low, but mounted high" (l. 172).[37] Indeed, Milton keeps experimenting with Spenser's revisions: reworking Spenser's play on Colin's invisibility, Milton begs a question at the very start of his poem: "Who would *not* sing for *Lycidas*?" (l. 10, my emphasis). Milton's alterations of the scene at Acidale are more thoroughgoing still, for it is the loss of Lycidas—not his intrusion—which disrupts the graces' ring:

> O the heavy change, now thou art gone,
> Now thou art gone, and never must return!
> Thee Shepherd, thee the Woods, and desert Caves,
> With wild Thyme and the gadding Vine o'ergrown,
> And all their echoes mourn. (ll. 37–41)

The nymphs who had shyly disappeared with Calidore's rude entrance are now blamed by Milton for departing (l. 50). But Spenser ends his career and the second installment of *The Faerie Queene* with a picture of knightly failure in the image of the Blatant Beast's "outragious spoile" (6.12.22. 9). This is a painful reminder that the work of poets, the legacy of Sidney, might even decay into rumor, as when the "tongues of mortall men, / which spake reprochfully, not caring where nor when," spare not "the gentle Poets rime" (6.12.27. 8–9; 40. 8). In this poignant picture Spenser gives us a sublime and solitary Sidney, his courteous example a rebuke, his gorgeous music unheard.

NOTES

Introduction "A nobility never interrupted"

1. See Diefendorf 661.
2. See Waller's (1993) discussion 22.
3. See Waller (1993) 20.
4. See Waller's (1993) revision of Freud's model of the family romance (16–51). Waller claims, for one thing, that worldly frustrations drive family members reluctantly back into the family circle: the "unrealizability of infantile desires . . . forces us back into the family as the site where we strive, inevitably imperfectly, to realize some equivalent for them" (39).
5. Stone (1977) suggests that the extended model remained more or less intact until 1550, and even then remained in slightly altered form, until the rise of "affective individualism" in the seventeenth century (4–5). For a discussion of how contemporary painting reflected this transformation of feeling, see Hopkins.

   Belsey proposes that Renaissance family life offered a new model of perfection to replace the discarded ideal of monastic life (xiii). Historians like Laslett have suggested important qualifications about changes in family life in the early modern period. See Laslett's challenge to "the classical form of Western nostalgia, the belief that families of the past were large, extended, patriarchal" in Laslett and Wall 8. See also Cressy 65–67; and Hill.
6. See Goody and also Perry for a related discussion of the privileging of the conjugal family by legal rules governing inheritance in the early eighteenth century.
7. My reading of the family as a discrete if permeable structure is offered in contrast to many new historicist readings, which, according to Boose, "[take] up the 'family' as a topic, only to then redefine it as the locus upon which the political state built *its* power through strategic appropriation, marginalization, and transformation of the family into an instrument of state authority" (731). That the family "has been the main agent of enculturation in Western history" is persuasively explored by Waller (1993) 30, passim. I would qualify Waller's findings in two ways, however: first, by "enculturation" Waller seems over and over again to mean gendering; and second, Waller focuses more on the shaping power of infantile desires and cultural pressures than on the mature efforts of self-consciously literary figures to transform such desires or resist those pressures.

8. Goody 4; and Medick and Sabean 10–11.
9. See Waller's (1991) discussions in "The Sidney Family Romance: Gender Construction in Early Modern England" (38) and in "Mother/Son, Father/Daughter, Brother/Sister, Cousins: The Sidney Family Romance."
10. *The Collected Works of Mary Sidney Herbert, Countess of Pembroke. Volume 1. Poems, Translations and Correspondence* 291. All references to Herbert's work will be to this edition.
11. Natalie Zemon Davis 87, 92.
12. As Natalie Zemon Davis comments, "it was not always clear where the nuclear family began and ended" (87). For an investigation of Shakespeare's handling of the early modern family, see Barber, who argues that "Shakespeare's art is distinguished by the intensity of its investment in the human family, and especially in the continuity of the family across generations" (188).
13. Stone (1977) 30, quoted by Hill 456.
14. For treatments of their relation, see McCoy and Quilligan (1988).
15. Sidney makes this claim in *The Defence of the Earl of Leicester* (1584), the only one of his works intended for publication.
16. A useful reading of the ways favorite sons were often indulged is offered by Helgerson (1976).
17. Wynne-Davies (2000) describes similar activities when she points to the Sidney's "family culture" and "Sidneian discourse" in her examination of the women of the Sidney family (167). At times, however, Wynne-Davies's picture of family life appears naive or strangely insensitive to its burdens: she claims, e.g., that "the perfect example of the 'safe house' concept may be seen in the Sidney family and their various residences: Philip's ideal knighthood, Robert's careful neo-Platonism, Mary Sidney's pious scholasticism, Mary Wroth's innovative independence and William Herbert's worldly statecraft are set off perfectly by the lauded pastoralism of Penshurst, the learned 'academy' of Wilton, and the clandestine passage of Bayard's Castle. Here the social standing, wealth, and cultural position of the family combine with the beautiful, gracious, and protected houses, simultaneously propagating a discourse which enabled both male and female creativity" (165).
18. Poster claims that before the appearance of the Oedipus complex, "father and mother exist only in pieces and are never organized into a figure nor into a structure capable at once of representing the unconscious" (106). Poster's work is cited both by Medick and Sabean 17–18, 25–26 and by Waller (1991) 407.
19. Terms for family relations often erase family borders. "Brothers" include stepbrothers, brothers-in-law, members of the same church (see Cressy 66–67); "uncle" could mean any older male relative, "nephew" any younger one. "Universal nephew" is Katherine Duncan-Jones's (1991) term for the much-admired Philip Sidney. Laslett (1980) describes bastards as "spurious children" in Laslett, Oosterveen, and Smith (217–46).
20. Orlin describes the familial rhetoric of Elizabeth Tudor: "Her tropic consanguinities adopted the characteristics of their familiar models, especially in

their uneasy yoking of affect and hierarchy" (85–86). But Elizabeth also dismantled these familiar models; the queen, Orlin claims, "exploited her very loss of patriarchal and sibling relationships rhetorically, by insinuating and manipulating fictional familial relationships" (85).

21. See also Medick and Sabean 9–27.
22. See Helgerson (1976); Guillory; McCoy; and Falco.
23. Lipking viii.
24. See Stone (1977) 4–5, 111.
25. Stone (1977) 7. Elsewhere Stone (1977) claims that the premodern family was "transient and temporary," "short-lived and unstable," "a low-keyed and undemanding institution which could therefore weather this instability with relative ease" (55, 60).
26. Lipking describes the "consistent internal logic" governing "the life of the poet" (viii).
27. I take the term from Stone (1977) 103.
28. See Miller (1996); Lamb (1990), Hannay (1991); Hackett (2000); Brennan (2000).
29. See note 17 and Wynne-Davies, " 'So Much Worth' " (80). My reading of this circle, its extent and secrets, challenges Wynne-Davies's suggestion that Penshurst was a secure and supportive setting for female artistry.
30. See Lamb (2001) 108.
31. See Quilligan (1989).
32. Gay 16.
33. Barber 192.
34. See Godzich and Kittay; and also Fowler and Greene.
35. Bloch 112. Lamb directed me to Bloch's work.
36. Cavanagh treats Wroth's relation to the intellectual milieu of the seventeenth century and Laroche examines Wroth as a contemporary of Shakespeare. See also Quilligan (1997).
37. Lamb (2001) links the romance of the *Urania* with both the Sidney family genealogy and broader generic concerns: "In various ways, Wroth's *Urania* represents an early modern continuation of an evolution traced through French medieval romance by Bloch, who demonstrates a profound relationship between 'genealogical succession and narrative structure' " (108).

Chapter 1   "Natural" Boys and "Hard" Stepmothers

1. Freud, "Inhibitions, Symptoms and Anxiety" (1926) 137; quoted by Klein (1975) 125.
2. *The Poems of Sir Philip Sidney*, ed. Ringler (1962). All references to Sidney's poems are to this edition and will be cited parenthetically in the text.
3. Klein 123.
4. See Winnicott (1957) 144. Krier's work directed me to Winnicott, although she and I draw on his thinking with different ideas about what Krier calls the "conditions of infancy" (295).
5. Montrose (1983) quotes Barber's claim that "the very central and problematical role of women in Shakespeare—and in Elizabethan drama generally—reflects

the fact that Protestantism did away with the cult of the Virgin Mary. It meant the loss of ritual resource for dealing with the internal residues in all of us of the once all-powerful and all-inclusive mother." As Montrose adds, however, "a concerted effort was in fact made to appropriate the symbolism and the affective power of the suppressed Marian cult. . . . The Queen was the source of her subjects' social sustenance, the fount of all preferments" (63–64).

6. Neale claims this speech was known to posterity as "the Golden Speech of Queen Elizabeth" (391). The speech is cited by Coch, who explores the biological and political pressures weighing on Elizabeth to produce a successor, along with "the range of meanings that 'mother' possessed in the [early modern] cultural imagination" (450). Montrose (1983) also mentions Elizabeth's speech (78).

   Compare the queen's address (delivered two years before her death) with her 1563 speech, presented so much earlier in her biological and political careers, and designed to address anxieties about the royal succession after one of Elizabeth's illnesses. The queen forestalls the marriage question out of "consideration of my own safety": "For I know that this matter toucheth me much nearer than it doth you all, who, if the worst happen, can lose but your bodies; but if I take not the convenient care that it behoveth me to have therein, I hazard to lose both body and soul" (Neale 108). Elizabeth acknowledges the queen's "two bodies" only to construe motherhood as an entirely political project: "And so I assure you all that, though after my death you may have many stepdames, yet shall you never have any a more mother then I mean to be unto you all" (Neale 109).

7. As Coch puts it, "[b]y transforming motherhood into a self-descriptive metaphor, Elizabeth escaped practical and cultural constraints while retaining the political benefits of an emergent maternal authority" (425).

8. Winnicott (1962) 15.

9. See Winnicott (1962) 18.

10. For a brief discussion of Stella's handiwork see Hackett (1998) 170.

11. Quilligan (1988) reads the connections between Stella and Penelope Rich (and Lettice Knollys) from a slightly different angle (186).

12. See Esler, and Helgerson (1976) for two accounts of the way such conflicts clouded political and literary ambitions alike.

13. Bateson 208, 215. Bateson's picture of the "double bind" relies heavily upon the image of the hostile, insecure mother: "If the child correctly discriminates [the mother's] metacommunicative signals, he would have to face the fact that she both doesn't want him and is deceiving him by her loving behavior. He would be 'punished' for learning to discriminate orders of messages accurately. . . . [Or] he must deceive himself . . . in order to support mother in her deception" (214). While I reject the explicit sexism of Bateson's account, some of the broader connections recent critics have drawn between mothers and children with Elizabeth I and her courtiers make his model nonetheless useful.

14. Privilege and privation are linked as early as 1398 in the loaded meaning ascribed to "private" (*OED* 4a): "Not open to the public; restricted or intended only for the use or enjoyment of particular and privileged persons."

15. For background information on the Renaissance development of the private sphere, see Fumerton, and Mazzola.

16. Krier, e.g., describes Spenser's innovations in the "limited register" of Petrarchism. An account of the apology as an "infantilizing genre" is provided by Ferguson (1983). See also Joseph Loewenstein (1985), who claims that "Astrophil's self-representation regularly splits him between the schoolroom and the nursery, and often makes sites of disappointment of both." "Sidney's Truant Pen" (134, 136). For the younger child, of course, these two sites would be identical. As is obvious to any reader of this chapter, Loewenstein's work has been of tremendous importance to my thinking, particularly his comments on a "late Elizabethan tradition of literary meditations on youth culture, a tradition in which *The Shepheardes Calender* as well as city comedy participate" (141–42).

17. Bateson 214.

18. Bateson 202–03.

19. For an account of the reciprocal nature of this process, see Montrose (1985).

20. I borrow this term from Joseph Loewenstein, who describes the psychological costs exacted by the ambiguities of the Elizabethan political scene (130). McCoy likewise argues of the *Arcadia* and *Astrophil and Stella* that "Sidney's rhetoric is alternately passive and aggressive, defensive and impulsive" (61).

21. Winnicott (1957) describes the "survival of the mother" in face of infantile aggression (144). Winnicott's account has a limited application to Renaissance culture, where wetnurses more commonly tended well-born babies. But see Adelman (1992) for an argument about the suspicions wetnursing aroused in Shakespeare's England (4–5).

22. See Joseph Loewenstein 133. Shearman similarly suggests the "endless opportunities" for boredom in a courtier's life. I am grateful to Jon Quitslund for directing me to this source.

23. For analysis of Sidney's New Year's gift as evidence of his "sexual subjection" see Minogue (555). Quilligan (1988) briefly takes up the episode (180) as does Joseph Loewenstein (132).

24. "Instructions for A.B." [Elizabeth's letter to Sir Francis Walsingham] Cecil-Stamford-Towneley MS (396–97).

25. Thomas (207). Thomas's work is cited by Joseph Loewenstein (133).

26. Thomas 212 n1.

27. Greene (1991) 3.

28. See Kay 34. Kay also relays Ben Jonson's story about Sidney's mother, who had contracted a disfiguring case of smallpox while caring for the queen, to suggest yet again how Elizabeth's "beauty was bought dear, and [how] the cost . . . fell to others" (23).

29. Berry suggests that English poetry under Elizabeth suffered a steep decline at the beginning of her reign, and directs readers to May's statistics about literary output during this period in *The Elizabethan Courtier Poets: The Poems and Their Contexts*.

30. See Duncan-Jones (1980) 161. McCoy similarly notes of Sidney's output: "All the major works are marked by inconclusive development, thematic

contradictions, and problems of closure" and they all "culminate in an impasse" (26).

31. Krier 304–05.

32. Thomas likewise suggests the new parameters of a world with so few members on top: "So far as the young were concerned, the sixteenth and seventeenth centuries are conspicuous for a sustained drive to subordinate persons in their teens and early twenties and to delay their equal participation in the adult world. This drive is reflected in the wider dissemination of apprenticeship; in the involvement of many more children in formal education; and in a variety of measures to prolong the period of legal and social infancy" (214). This passage is cited by Joseph Loewenstein 133. See also Lamb (1994), who explores how Latin tutors were viewed as having rescued young boys from a female culture of nursery rhymes and vernacular fictions (504–05).

33. I take the term "foster child" from the triumph Elizabeth staged as entertainment for the French ambassadors representing Alencon. In it, the "four Foster Children of Desire" besieged the Castle of Perfect Beauty only to be defeated by the defenders of Beauty's fortress. As Joseph Loewenstein reminds us in his analysis of the episode, Sidney was one of the four foster children (131). "Stepson" is a much looser allusion. As son of Henry Sidney, charged with the government of Ireland, and heir to Leicester, the Queen's favorite "favorite," Sidney would have been seen by his contemporaries as enjoying a quasi-familial, quasi-legal connection to Elizabeth.

34. Cf. Kalstone, whose reading emphasizes the architectural rather than destructive impulses of Sonnet 71 (117–18).

35. For examples of this critical uncertainty, see Rudenstine 12; and Joseph Loewenstein, who maintains that Sidney's is a poetics of self-inflicted humiliation (133, 141). Sidney's mentor Languet, in his letter of October 22, 1578, combines Sidney's tendencies toward abasement with those toward pleasure, reproaching Sidney for trying to "escape the tempest of affairs" by retiring to the "privacy of secluded spaces."

36. I borrow the phrase from Waller (1985) 239.

37. Krier likewise describes the discomfort with Petrarchan sentiments and strictures in the Tudor period, pointing to Spenser's remedies in the *Amoretti* and *Epithalamion*, where the "diction, motifs, and themes [which inform] these poems are inaccessible, even unimaginable, to the tormented and exasperated Petrarchan lover" (308).

38. Quilligan (1988) 188. Elsewhere Quilligan (1988) describes "the underlying politics of the sequence as a subtle resistance to the queen" (190) and points out that "Sidney addressed no poem to Elizabeth, save for *The Lady of May*" (195 n35).

39. See Archer 53.

40. See Duncan-Jones (1989) edition of Sidney's works (viii).

41. See *The Prose Works of Fulke Greville, Lord Brooke* (37–41). Hereafter, references to Greville's text will be made parenthetically in the text.

42. Sidney, *The Defence of Poetry*, ed. van Dorsten (18); hereafter cited in the notes as *DP*. Lamb (1994) explains that, "[l]ike rabbits and ducks, warriors

and infants perform a mental *trompe d'oeil* as they vie for the prominent position in the *Apology* which will render the loser invisible" (500–01). In analyzing Sidney's motives, Lamb also points to early modern fears that the *effeminizing* pleasures of poetry might reduce the male poet to immaturity or "the essential femininity of early modern boys"; but I think Lamb's reading overlooks the kind of power that comes with childishness. See also Joseph Loewenstein 137 and McCoy 76–84.

43. *DP* 31.
44. *DP* 34.
45. *DP* 40.
46. *DP* 18. Lamb (1994) cites Joseph Loewenstein's findings, "drawn from the many references in *Astrophil and Stella* to the schoolroom and the nursery, that poetry was perceived as essentially childish" (514). I would argue that it is not so much that poetry is childish as that the Elizabethan poet had to operate as a child. Along these lines, Berry describes Sidney's "continuing vocational crisis," and says of *The Defence* that "the character of the persona [is] . . . itself a persuasive argument" (x, 4).
47. *DP* 18.
48. *DP* 61.
49. Joseph Loewenstein hints at such a reading too, commenting that Sidney figures "desire itself as regressive. . . . [Astrophil] is plainly a novice, animated by adolescent jealousy and adolescent disdain for those who have subdued courtship to routine" (134).
50. See Marotti; and also Crewe's objections to Marotti's "antiromance" model (72–76). According to Crewe, Marotti insists on the "assimilation of major sixteenth-century authors to limited patterns of competition, clientage and career advancement," "indeed to limited conceptions of the political as such" (72). In some ways this chapter endeavors to broaden such conceptions and propose that politics makes up a good deal of a poet's earlier life.
51. Greville's is the only contemporary account of the tennis court disagreement between Sidney and the Earl of Oxford. Duncan-Jones (1991) connects this episode with Sidney's letter to Elizabeth (163–64).
52. Quilligan's (1988) reading of this episode privileges the sociology of Bourdieu over Huizinga, making aristocratic play (and provocation) a matter of exchange rather than of rigorous, even scientific experimentation (172). Both contestants are remade in Greville's account, according to Quilligan's analysis.
53. See Quilligan (1988) 173. She claims that in both his argument with Oxford and in his sonnet sequence Sidney takes control of his inferior social situation (171), putting Oxford in his place by insisting upon Oxford's position above Sidney. Moreover, she proposes that "Sidney's humility—and specifically his sense of his own socially inferior position—becomes a weapon of honor against the very hierarchy that would limit the power of the inferior position" (172). I think Quilligan's shrewd analysis of this episode and of Sidney's continued criticisms of the queen's nuptial choices nonetheless gives the poet too much credit and Elizabeth too little. Quilligan maintains, e.g., that: "Humility and a profound sense of social inferiority allow Sidney not

merely to triumph over, but to obliterate, Elizabeth's challenge to his authority; such a move becomes a virtual Sidney signature" (177). This interpretation transforms Sidney's lifelong arrogance into virtue and makes youthful
brashness a sign of wisdom. I would argue that Sidney's faults are just as
interesting and formative as his virtues. Much like the favorable spin
Quilligan puts on Sidney's argument with Oxford, Lamb (1994) locates
success in Sidneian circumlocutions such as the poet's attempts to surpass or
bypass the female role in reproduction in *The Defence of Poetry* (507).

54. The Pierpont Morgan Library, New York, has the autograph version (MA
1475). For a corrected and modernized text (the autograph appears rushed
and is nearly illegible in places), see *The Miscellaneous Prose of Sir Philip
Sidney*, ed. Duncan-Jones and Van Dorsten (1973), especially their introduction (123–28). Hereafter, their edition of this work will be cited in the
notes as *DL*.

55. *DL* 134.

56. Duncan-Jones (1991) takes this description of Sidney from Gloucester's
picture of Edmund in *Lear* 2.1.85 (269).

57. *DP* 24.

58. Rpt. in *The Prose Works of Sir Philip Sidney*, ed. Feuillerat (1962–69) 3:
166–67; quoted by Worden 46.

59. Hager (1991) argues for a different organizing principle, claiming that
"behind that angry mask lies an essay on slander and naming that develops
with wit a paradoxical theme concerning the effects of fame" (57).

60. *DL* 137.

61. *DL* 135.

62. *DL* 129.

63. *DL* 134–36.

64. Greenblatt (1985) 214. For a useful corrective to Greenblatt's romantic
picture of history, see Bellamy, who argues that "Greenblatt's argument
constitutes the paradigmatic new historicist one whereby psychic experience
disappears in the gaps of the subject's dispersal in the discursive formations
of ideology" (21).

65. See Duncan-Jones (1980) 164–65. McCoy takes Sidney's aspirations and
frustrations as both "distinctive and representative" and describes the "social
and personal predicament of an Elizabethan aristocrat, caught up in a
tangle of diminishing feudal power, Renaissance notions of statecraft, a zeal
for military glory, courtly dependence and intrigue, and a cult of devotion
to a formidable, emasculating queen" (x). Cf. Rudenstine, who calls Sidney
a "charming truant," and Joseph Loewenstein, whose picture of Sidney's
"masochism" suggests something darker in the shape of a self-confessed
criminal or Elizabethan public enemy (141).

66. See Duncan-Jones (1980) 162, 165.

67. See Buxton (1989) 46–47; and Berry 162. Esler describes the "honourable
recklessness" that killed Sidney (91).

68. See Berry 6–7.

69. Perhaps a similar signal—this time relayed by a representative of Elizabeth's
successor James—is sent when Ben Jonson reports that "Sir P. Sidney was

no pleasant man in countenance, his face being spoiled with pimples" (see Kay 19).

70. Quilligan (1988) also takes up this interview (174, 178).

Chapter 2   Brothers' Keepers and Philip's Siblings

1. Henry James's statement is quoted by Feinstein 301.
2. See Duncan-Jones's (1989) edition of *The Major Works* (viii); and also Duncan-Jones (1991). Additional adjustments to our picture of Sidney are proposed by Montrose (1977); Levy; and Lisa M. Klein, who argues: "While modern critics usually seek to heal or transcend the division [between Mars and Mercury] by either valorizing the poet over the Protestant man of action or, in the case of new historicists, the reverse, I maintain that the unresolved tension between these roles is essential to understanding the dynamism of Sidney's example" (39).
3. Tamen describes the "delicate task of controlling ghosts" as at the heart of most projects to shape a national literature or national identity (295–97).
4. As Stone (1977) argues: "The family . . . was an open-ended, low-keyed, unemotional, authoritarian institution which served certain essential political, economic, sexual, procreative and nurturant purposes. It was also very short-lived . . . frequently dissolved by the death of husband or wife" (7).
5. Orlin explores many of the transmutations of familial bonds at this time. E.g., Elizabeth I referred to her successor James VI as "brother" even more frequently than as "son," demonstrating, as Orlin argues, that Elizabeth "exploited her very loss of parental and sibling relationships rhetorically, by insinuating and manipulating fictional familial relationships" (85). We might add that this kind of exploitation regularly characterized royal behavior: Mary Stuart's mother only formally recognized Mary's half-brother and sister after their father King James V's death; his own recognition surely would have carried emotional as well as political freight.
6. Rpt. by P.J. Croft, *The Poems of Robert Sidney* 42. Although I use Croft's numbering of Robert's poems, I direct the reader to Germaine Warkentin's important cautions about this edition.
7. Besides Warkentin's essay and Croft's edition of Robert's poems, other notable exceptions are James; the biography by Hay; the study by Kelliher and Duncan-Jones (1975); and Duncan-Jones's (1981) edition of "The Poems of Sir Robert Sidney" in *English*. Without referring to his poetry, Esler uses Robert to provide a "generational case study." See also Waller (1975–76).
8. For an illuminating examination of this phenomenon, see *Blood Brothers: Siblings as Writers* (note 1), a study of later writers, which, however, only examines brothers. Rather striking in this collection is the pervasive notion that shared creative abilities are somehow debilitating or divisive; the accounts are heavily influenced by psychoanalytic models and most of the sibling struggles appear as distorted Oedipal conflicts with physically damaging consequences. Examining sibling professionals in a related field, Wolfe's account of twin gynecologists appears to take this notion of sibling

debilitation to an extreme. Wolfe reports that the two doctor brothers secretly saw each other's patients, refused to perform surgical procedures without consulting the other, and encouraged each other's suicide. The brothers' extreme closeness, for Wolfe, requires taking up more primitive ideas about contagion (or sympathetic medicine?). Some of this debilitating closeness also seems to have characterized the brothers' erratic professional habits; given that the two coauthored a prominent gynecological textbook, their bizarre tie serves as a terrifying endpoint for many Petrarchan practices and obsessions.

9. I take the term from Wynne-Davies's (2000) essay in *This Double Voice*.

10. Hanson comments on the "asymmetry in the way gender [operates] in [Renaissance] literary production" (168). A useful study of families as sites of culture is provided by Wrigley (71–72). See also Waller's (1991) essay "Mother/Son, Father/Daughter, Brother/Sister, Cousins."

11. Wrigley suggests, in fact, that the Renaissance family was a prime locus for culture. Unlike other institutions, "commonly restricted by a single function or range of functions—political, economic, or religious," early modern family life "spanned virtually the whole range of human activities" and "[e]ffective membership of society at large was attainable in many circumstances only by membership of a family through which a claim could be mediated" (72). See also Wrightson.

12. See Waller (1993) 22.

13. *DP* 61.

14. See Waller (1993) 55.

15. John Donne, "Upon the translation of the Psalmes by Sir Philip Sydney, and the Countess of Pembroke his sister." Rpt. by Rathmell in his edition of *The Psalms of Sir Philip Sidney and the Countess of Pembroke* (ix–x).

16. For a fuller account, see Falco 7 and Duncan-Jones (1991) 194.

17. Danby describes how "four or five renaissance dynasties" "dominated" the scene, "almost all of them called into being and maintained in effective place by the creative *fiat* of the Tudor despot" (22).

18. Stone (1967) 61. As a result of these transformations, the Renaissance family had also become something of a *literary* structure, subject to interpretation and documentation, something also codified and symbolized since family members not only needed lands and titles but emblems and narratives to protect themselves from the challenge posed by a rising merchant class. Even Elizabeth I and her father endeavored to trace their tree to King Arthur, and on such slender branches and crucial stock Mary Stuart hoped to attach herself to her cousin.

19. See Worden for a shrewd discussion of the economic pressures weighing on Sidney and his father (43–44).

20. Stone (1977) describes the gradual breakdown of the older, relaxed family model ("The Open Lineage Family") beginning about 1530 and its replacement by a tighter nuclear model ("The Restricted Patriarchal Nuclear Family") with sharper allegiances to one's father, Church, and state. Stone claims this new type "predominated from about 1580 to 1640" (7). Williams describes a much later development epitomized by the transparencies of the

rural village and characterized by closely tied, constantly—if carelessly—monitored and economically aligned "direct relationships," like the one imagined in George Eliot's *Middlemarch* (165–81). As Williams indicates, all of the standard machinery—economic, familial, and political—of the "knowable community" are comfortably settled at aristocratic estates where pastoral acquires a specific "social base" (27–34).

21. See Williams and Wayne.

22. See Levy 7.

23. See Duncan-Jones (1989) for an insightful discussion of the discomforts of the adult Philip's position.

24. Hulse persuasively suggests that much of the difficulty in *Astrophil and Stella* is a deliberate effort on the part of the poet to prevent others from deriving meanings intended solely for Sidney's ideal reader: "[T]he historical audience, the ideal reader, and the principal reader described within the poems are all one person, Penelope Devereux Rich," making court wits and male readers only a cover for the real critic (273).

25. Wrigley 71–72.

26. This apparently had been a long-standing problem for, as Antonia Fraser tells us, "when the queen [of Scots] was a baby, the ideal thing might seem to be a second son of 'France, Denmark, or England if such a thing existed'" (241).

27. We would do well to remember Schwartz's caution against any easy familiarity with the term: "The difficulty any discussion of kinship must address is that, far from successfully escaping the artifice of identity, kinship systems are themselves artificial. After long and tortuous debates about the significance and forms of kinship systems, anthropologists are now telling us that there is virtually no such thing as kinship. There are *ideologies* of blood relations, *constructs* of brothers and sisters, but comparative studies have shown us how diversely such notions are understood. There are no real blood relations" (78).

28. In his 1580 letter to his brother, Philip Sidney indicates how much funding Robert can expect from Leicester, and promises that an additional amount from their father "shall come with the hundred, or else my father and I will jarl." *The Miscellaneous Works* (337).

29. Esler 54–55.

30. Esler 54.

31. Greenblatt (1980).

32. Machiavelli 14.

33. As Schwartz and Shuger have similarly argued, the Protestant recovery of the Hebrew Bible recovered Hebrew family systems too, and early modern issues surrounding kinship became loaded with their political, dynastic, interpretive, and theological weight (see Shuger 40–41). Shuger describes the eruptions of filial defiance during the early stages of the Reformation when sons and daughters converted to Protestantism (153). Many families were disrupted in England almost at the outset: when Tyndale introduces his English Bible in 1530, torturous debate sparks over questions like the legality of Henry VIII's marriage to his brother's wife and, later, over the codes of exogamy establishing whether Anne Boleyn was Elizabeth's sister

or mother. The Tudor family was primarily understood as a legal structure and secondarily as polite fiction, organized by jurists, less tried by affection.

34. See Schwartz's introduction.

35. Waller (1993) borrows the term from Julia Kristeva (155–56, 284).

36. Falco reminds us that Spenser, only two years younger than Philip Sidney, is habitually viewed as belonging to the next generation, notwithstanding the 1579 appearance of Spenser's *Shepheardes Calender* (53–55).

37. See Waller (1993), especially chapter 1: "The Sidney Family Romance: Breeching the Subject."

38. Waller (1993) cites John Aubrey's notorious speculations. See also Roberts's introduction to *The Poems of Lady Mary Wroth*.

39. If Waller faults Mary for composing no original work, others credit her with the myth surrounding her brother. See Hannay (1990).

40. James 185.

41. See Waller (1990) 337.

42. See *Miscellaneous Works* for Philip's two letters to his brother (329–39). The first letter, probably composed in 1578, recommends Aristotle's *Ethics* and encourages the development of "the knowledge of such things as may be serviceable for your country and calling." Homer appears here as a moral philosopher (329–31). The famous second letter of October 1580 supplies "sweet Robin" with a list of readings, including Bodin, Tacitus, Livy, and Plutarch as examples of logicians.

43. Cf. Buxton (1964), who compares Sidney as well as his successor Essex to Spenser's knight of courtesy Calidore (207).

44. See discussions by Falco 2–3 and Guillory.

45. See Guillory vi; and Helgerson (1976).

46. See Falco 2, 12.

47. Duncan-Jones (1980) claims that all of Philip Sidney's literary efforts were immature "toys," and that he was never able to complete the serious work he so desired to undertake.

48. This kind of shielding mechanism or "secret service" might also be tied to Philip Sidney's father-in-law Francis Walsingham's courtly efforts as head of the Elizabethan secret service. See studies by Archer and Plowden, although neither makes an explicit connection between Sidney and Walsingham.

49. See Croft's edition and Duncan-Jones (1981) for this dating of Robert's work (xiv). Fisken supplies the dates of Mary Sidney's efforts (273 n6). Additional information is supplied by Roberts (1983) and Waller (1975–76) 694. With a characteristic mix of prescience and hindsight Waller (1975–76) comments that Robert "may have decided that Mary, more permanently settled at Wilton in the 1580s with the increasing comings and goings of Greville, Spenser, Daniel, and other poets, was better placed to direct the Sidneian literary revolution" (694).

50. Greene (1991) 3 and a (1990) discussion of "Sir Philip Sidney's *Psalms*."

51. I here use Duncan-Jones's (1989) edition of Sidney's poetry.

52. Cf. Greene (1991), who makes Astrophil rather than Stella the source of poetry: Astrophil "proposes that [his rivals] start their works again from infancy, and follow his influence" (86–87).

53. See Greene (1990) 19.
54. Greene (1990) 26.
55. Greene (1991) 17.
56. There are many variants of the Sidneian psalms in both quantitative verse forms and simpler ballad stanzas. I rely on Rathmell's edition of the completed *Psalmes*, although one might agree with Waller (1990) that no final version of the Sidneian psalms really exists. Characterizing Mary's experiments in terms of "obsessive repetitiveness," Waller suggests this unfinished state reflects her uneasy position, "caught amongst conflicts over which she had little control and in which she struggled to 'own' a voice" (339, 338). I would argue that her variations reflect repeated confidence in a vehicle perfectly suited to her talents and interests. Elsewhere Waller (1979) writes that Mary "directed the flexibility, energy, and poetic suggestiveness of the new Elizabethan poetry into the religious lyric" (226). Cf. Goldberg's (1996) criticism of Waller's stinting judgment in "The Countess of Pembroke's literal translation": "The sign of propriety . . . is precisely what is denied Mary Sidney" (322).
57. See Hannay (1994) 45.
58. See Smith 264–65. Smith provides a very useful introduction to English psalmody, which I draw on extensively here. Many psalm versions, to quote Thomas Fuller, were written by "men whose piety was better than their poetry" (Smith 250). Still, Wyatt published a version of the seven penitential psalms (based on Aretino's prose version) in 1549, and Surrey followed with four translations while he was imprisoned in the Tower. But it was the success of Sternhold and Hopkins's 1562 collection—which used common meter throughout—that made psalmody part of popular culture and prompted more than six hundred editions (see Hannay [1994], Rathmell xiii). All in all, there were more than a hundred different psalm translations which appeared between 1500 and 1640.
59. Most readers date Philip's psalm translations to 1583 or 1584, although Spencer (1986) speculates that they are among his first literary efforts, "Sidney's earliest most important poetic task" (34). That Philip instead undertook these translations during the last year of his life is suggested by Seth Weiner (1986) 193.
60. See note 56. I employ the standard edition edited by Rathmell, which relies on the countess's final revisions of her brother's poems.
61. Rathmell discusses this psalm in his introduction, where he notes: "The image of the stillborn embryo has an immediacy that is certainly not present in the formal metaphor of the 'untimely frute' that we find in both the Geneva and the Bishops' Bible" (xxi). Some of Mary's verbal expertise comes from an apparently stronger linguistic background as well as, Waller (1974) suggests, the availability of Philip's work as a model. Models for Philip included Coverdale's edition of *Goostly Psalmes* (ca. 1535), Wyatt's verses, and continental models like the Marot/Beza psalter (see Rathmell xvi; Smith 269). Hannay (1994) maintains that Mary also had the example of earlier and sometimes unpublished psalm translations prepared by Protestant women like Anne Askew (50); see also Hannay (2000) and

(2001). Steinberg (1995) argues that Mary had the original Hebrew mean-ings within reach; he claims that much of what is explained by critics like Waller and Hannay as reflecting a Calvinist influence can be described as "reflecting a familiarity with the Hebrew originals"—if not firsthand linguistic knowledge then at least an acute awareness of the ancient histor-ical background (7).

62. See Hannay (1994).
63. Waller (1984) 69–70.
64. See Mazzaro 73–74.
65. Waller (1975–76) supplies some of this background information (691).
66. See Waller (1993) 61. He describes the intimacy between Mary Wroth and William Herbert as a "family romance centered on the substitute father figure, who possessed some authority, but also posed no threat" (76).
67. See Tamen 295–97.
68. See also Waller (1993) 102.
69. See Roberts's introduction to Wroth's poems (37).
70. For other discussions of Mary Sidney's importance to seventeenth-century poetics and beyond, see Steinberg, Smith, and Waller (1979). Rathmell maintains that a serious consideration of Mary's work suggests "the history of the metaphysical revival of our own time . . . be rewritten." For more extended reflections on the importance of English psalmody to English verse see Campbell, who argues that the history of the lyric in England is really a history of psalmody.

Chapter 3   The Grammar of Families in Sidney's Old *Arcadia*

1. Milton's *Doctrine and Discipline of Divorce* will be abbreviated *DDD*. See Patterson's discussion (85). Although Patterson does not comment on Sidney, I find her discussion enormously compelling and instructive for a reading of the *Arcadia*, as I do her comments on Milton's experiments with romance and on romance's "novelistic future" (85–86). An excellent treat-ment of the *Arcadia* and of Renaissance romance fiction more generally is provided by Newcomb (2002).
2. As punishment for her crimes against the state and family, Sidney will later bury Gynecia alongside her presumably dead husband.
3. A number of critics have anticipated many of these concerns with the family and narrative without bringing them together and I draw heavily on their work. See, e.g., Bono; Cantar; Martin; and Tennenhouse's (1986) investigation of the connections between Sidney's letter to Elizabeth and the *Arcadia*, rpt. as "Arcadian Rhetoric" in *Sir Philip Sidney's Achievements*. As Quilligan (1989) comments, "the *Arcadia* in effect continues Sidney's consideration of the problematic Tudor traffic in women, begun so inaus-piciously for his court career in 'The Lady of May,' and continuing in the infamous 'Letter concerning Monsieur'" (267–68). Stillman (1985) chal-lenges any linkage between the two texts, however: "Why Sidney should need to cover his opposition to a marriage in fiction which he so openly argues against in his letter to the Queen in not clear" (799–800).

4. Many critics have adopted the phrase "mixed mode" to describe Sidney's experiments with genre in the *Arcadia*, where he combines pastoral verse with epic heroics, romance intrigue, domestic comedy, and philosophical irony. See Walter Davis, and Greenblatt (1973), who calls *Arcadia* "perhaps the supreme Elizabethan example of what I shall call the mixed mode . . . that strange conjunction of literary kinds" where "to resolve is to lie." Greenblatt, "Sidney's *Arcadia* and the Mixed Mode" (269, 278).

   By focusing on Sidney's narrative as an example of early modern prose more like than unlike our own, my claims differ from those of Adolph, who sets out as a general requirement for modern (i.e., seventeenth century) prose that it "be a vehicle of useful communication rather than a medium which calls attention to itself either as conscious art or self-expression" (6–7). Still, I am indebted to Adolph's thinking, as I am to other studies of Renaissance prose, which, though relatively few in number, are rich in scope. See Croll's work, collected in *Style, Rhetoric, and Rhythm*, and Fowler and Greene's essay collection, which advances and extends Croll's influential agenda.

5. All references to Sidney's Old *Arcadia* (hereafter abbreviated *OA*) are to Duncan-Jones's (1985) edition, which follows closely the edition prepared by Jean Robertson (1973). Page numbers to Duncan-Jones's edition will be listed parenthetically in the text.

6. Tennenhouse (1986) 201.

7. I am indebted to Martin's discussion (369).

8. Martin 369.

9. Cf. Dipple (1971), who says of Pyrocles's transformation: "The mythic implications that Sidney subtly suggests . . . thus inform that reader whose mind is as active in the detail of both structure and meaning as Sidney's style demands, that Pyrocles' choice is of [Daphne's] virtue and that Musidorus's arguments [against Pyrocles' transformation] are prohibitive of that virtue. Nevertheless, Pyrocles is not Daphne, Musidorus is not the lustful Apollo, and the analogy, for all its implications, is finally as amusing as it is suggestive" (51). These patterns of metamorphosis are revised in the New *Arcadia* where, Dipple explains, "what had seemed comically feminine and sexual in the *Old Arcadia* is . . . made martial and masculine. Even more significant, the whole idea of metamorphosis takes on a new cast. Instead of assuming total physical change which is urged by a new interior experience and performed with idealistic ignorance, . . . Pyrocles' old self is incorporated into the new self and . . . he is not launched completely into a new persona" (57). See also Altman, who describes the narrated adventures of the princes as "heuristic exempla," which suggest "not a temporal development so much as a pattern of exhibition." Dorus himself reflects on one of these episodes: "This adventure, though not so notable for any great effect . . . [is] yet worthy to be remembered for the un-used examples therein" (90, 92).

10. See Godzich and Kittay 8.

11. Rudenstine reads Arcadian mirroring differently, suggesting "Pyrocles, unlike Sidney in 1578–79, is a truant from the noble school of heroic virtues in which he had been bred" (19). Elsewhere, however, Rudenstine

calls Pyrocles an "unofficial apologist for Sidney" in that he is "an unwitting apologist for poetry" (23). Similarly, Hager (1991) describes Sidney as "a master of the overview that his exemplary image would deny him. In his poetry, criticism, and fiction, Sidney presents us with a sequence of over-reachers who are undercut or trapped by rhetorical and narrative irony, but these glorious over-reachers are all self images" (20).

12. Godzich and Kittay note that by convention literary works are *either* prose *or* verse, because "[e]xpression admits of no other possible forms" (ix).

13. Stillman's (1986) work on the dialectical or agonistic relation of poetry and prose in the *Arcadia* remains the most useful study for my own argument, although Stillman limits the problem to a literary rather than existential one for Sidney.

14. Cf. Preston, who suggests that this is the case especially in Sidney's New *Arcadia*: "The spectacular increase in ekphrastic material [is] one of the most noticeable features of Sidney's revision of the romance." "This increase," Preston observes, "produces a fictional universe of legible pictures in which all artificial and natural phenomena have potential 'countenance'" (97). See also Olmsted who, focusing on the New *Arcadia*, argues that Sidney uses tropes of disguise to explore changes in class structure, in the social roles of gentry and aristocrats, and in the institution of marriage (172–73). Olmsted further explains that Sidney "engages poetically with the opportunities and problems introduced by the increased social mobility and by the shift in the definition of the social role of the nobleman from one of warrior to one of educated bureaucrat and advisor" (177). I also draw on Clare Kinney's unpublished essay "Hybrid Text or Hypertext?": Mary Wroth's Recombinant Poetics of Genre," which explores how the *Urania*'s "composite form may itself reconfigure the signifying practices of both the lyric sequence and the romance."

15. Barkan is quoted by Olmsted 167.

16. Of course, this stipulation is not entirely a product of the Renaissance. As Godzich and Kittay remind us, "medieval biblical exegesis recognized the necessity of applying a set of different literacies, four 'levels,' to one text: the literal, the tropological, the allegorical, and the anagogical" (112). Such a talent, I would argue, might have previously—and profitably—been divided between sinner and saved or, in Chaucer's universe, between cleric and wife. The Renaissance innovation requires *everyone* to possess different literacies.

17. Lamb (1997) argues "Musidorus's affirmation of his 'real' aristocratic nature speaks to Sidney's problematic location in class in the 1580s" (58). Cf. Prendergast, who emphasizes Sidney's medium over his message, suggesting "Sidney was attempting to determine his own place in the transition from an aural to a print culture" (102). Hager (1991) places Sidney's efforts within a still broader context, claiming that the poet "helped usher in one of the supreme eras of the stage and of fictional characterization in general" (7).

18. Altman 91. See also Maslen 285–88, where he explains the preoccupation of many Elizabethan writers with narrative subtleties as derived from an

anxiety about the act of producing fiction. "Like Gascoigne and Lyly," Maslen argues, "Sidney depicts his heroes as spies, and makes his narrator complicit with their subterfuges" (296). Sidney's case is unique, however: "The relative impunity afforded him by his aristocratic status gave Sidney the opportunity to carry the word-games of his predecessors to their terrifying conclusion: to convert their petty treasons to high treason" (297).

19. See Patterson 86.
20. See Worden 11.
21. Here I draw on Patterson's exploration of Milton's "syntactic symptoms" (94).
22. I should note that Greene (1997) examines how Sidney's work overall offers a "theory of fiction as embassy" (177–78).
23. Sidney, "A Letter Written by Sir Philip Sidney to Queen Elizabeth, Touching Her Marriage with Monsieur" in *Miscellaneous Prose*, hereafter abbreviated as "A Letter."
24. Spencer (1945) 268.
25. Spencer (1945) 266.
26. Cited by Maslen 294.
27. Sidney's story of Amasis, son and heir to Sesostris, king of Egypt, offers a brief recapitulation of this problem. As Histor relates, "the king Sesostris, after the death of Amasis's mother, had married a young woman who had turned the ordinary course of a stepmother's hate to so unbridled a love towards her husband's son Amasis that neither the name of a father in him, of a husband in her, nor of a mother and son between themselves, could keep her back from disorderly seeking that of Amasis which is a wickedness to accept" (*OA* 137–38).
28. Greene (1997) 180.
29. See Muir's introduction to *Elizabethan and Jacobean Prose* (xxi).
30. See Godzich and Kittay's discussion of modern prejudices (x).
31. See Fowler and Greene's introduction (1).
32. See Fowler and Greene's introduction; and Hamilton, who comments on the typical neglect of Elizabethan prose fiction; "[t]o its misfortune," he notes, "Elizabethan prose fiction was succeeded by the novel" (22–23).
33. See McKerrow's introduction to *The Works of Thomas Nashe*, 5 volumes (London, 1904–10), quoted by Adolph (271–72).
34. Imbrie 46–47.
35. *Versiprosa* texts are also known as *prosametrum* (see Godzich and Kittay 46).
36. See Prendergast 99. In a footnote, Prendergast qualifies the rather stark opposition she draws and explains her decision "not to dwell on in-between or heterogenous modes of narrative in [the *Arcadia*]": "Such utterances as inset epistolary narrations, poems expressed outside of the eclogues, or the story of Plangus told within the eclogue sections are most often placed on the margins of the prose and eclogue sections of the work" (99; 224 n4). I think Prendergast overstates her case for a "gender-based poetics that informs all three narrative levels—prose narration, the eclogues that conclude each section of the work, and the framing asides by the narrator to his audience of 'fair ladies' " (100). Indeed, Prendergast later qualifies her

position again, arguing that "Sidney shifts what was previously seen as two forms of verse—lyrical and narrative—to contrast between a static, idolatrous verse of the past ('Petrarchan') and fluid, allusive, emerging prose ('Ovidian')" (109).

Approaching the matter in more formalist terms, Stillman (1986) suggests that the Old *Arcadia* "is constructed out of divided and distinguished worlds," moving from "forward-moving" "individual" narrative to a separate and "static, emblematic" "universal" world of verse, where the eclogues function as "entertainments separate from the rest of the action" (81–82). Elsewhere Stillman (1985) notes : "as [Sidney] retreats from the prose books to the Eclogues, he comments upon the nature of his 'retreat' from the world of history to the world of fiction" (803).

37. See Donovan; Bahktin.
38. See Flint's investigation of the anthropology of the eighteenth-century family; and Donovan's discussion of "feminist prosaics." Mary Ellen Lamb shared a forthcoming essay on Mary Wroth, which explores her romance as something on the verge of collapse; in chapter 4 I argue that Wroth's shift to "modes of association" is what allows romance to persist, to become prosaic, to treat slighted younger brothers or even the children of her protagonists.
39. Godzich and Kittay 116.
40. For a highly interesting discussion of Wroth's sonnet sequence as a "proliferation of (subversive) equivocation," see Clare Kinney's unpublished essay "Hybrid Text or Hypertext?"
41. See Relihan (1994) 4.
42. Similarly, euphuism, in Carey's terms, derives from the "constant application of power to control an exploding environment" (364). Corthell offers a more sanguine treatment of Renaissance prosaics (7).
43. Cited by Relihan (1994) 4.
44. For discussions of how royal favors were redistributed, see Whigham; and Stone (1967).
45. See Olmsted 168 for an account of these different forms of metamorphosis. See also McKeon (2000), who claims that "what makes the novel a different sort of genre may therefore be not in its 'nature' but in its tendency to reflect on its nature—which of course alters it nature in the process" (4).
46. See discussions by Flint and by Spacks, whom Flint cites.
47. Archer supplies an important rationale for the distended shape of much Renaissance prose fiction, with More's *Utopia* an unlikely precursor to Nashe's *Unfortunate Traveller*: "The rationality that initially governed intrigue and diplomacy," Archer claims, "was different from the commercial rationality that eventually overtook it. It was not lucid, efficient, methodical, or quantitative. It was opaque, prudent, improvisatory, and obsessed with degree rather than number. It developed, in fact, from symbolic interaction among people of various degrees; it did not seek to reduce exchange to equivalence" (12).
48. That prose is a language filtered through itself rather than through court is suggested by Godzich and Kittay (202).

49. Prendergast 112.
50. Godzich and Kittay xix.
51. See Allcott's "General Introduction" to *Elizabethan and Jacobean Prose* (xiv).
52. See Muir's introduction to *Elizabethan and Jacobean Prose* (xvii). Readers of the *Arcadia* offer mixed accounts of its literary influence. Woolf writes in "The Countess of Pembroke's *Arcadia*" that in the "dream pages" of *Arcadia* "all the seeds of English fiction lie latent" (39, 48). Schlauch counters that "for all its literary virtues, we must admit that the *Arcadia* has less kinship with the modern novel than some of the late French prose romances which Caxton had put into robust English back in the 15th century" (185). See also Robinson 158–59. Following Adolph, Robinson sees the new science, particularly as embodied by the Royal Society of London For the Improving of Natural Knowledge, as figuring largely in the development of modern prose.
53. Morenco 251.
54. Greene (1997) 177.
55. "A Defence of the Earl of Leicester," *Miscellaneous Prose* (134–35).
56. "Defence of Leicester" 137.
57. Godzich and Kittay 17.
58. See Martin's discussion of how these episodes are linked (373–74). Martin persuasively ties Sidney's letter to the opening crisis of the Old *Arcadia*: "The old duke's anxiety over his royal estate is surely intertwined with a need to confirm his virility. His daughter's impending marriage and succession to the office he covets in large part determines the sonless monarch's course of action. . . . Against the escalating threat to offspring that ensues, moreover, Sidney poises a selfless ideal of procreative continuity and renewal celebrated in the original version's Third Eclogues" (370–71).

   Straddling what Patterson calls an "undrawn boundary between polemic and narrative," "pressing against the frontiers . . . that divided the romantic past from the novel (and novelistic future)," Sidney's letter to the queen, like Milton's pamphlets on divorce, "[hold] interest for genre theory because [they hover] on an undrawn boundary between polemic and narrative, a boundary whose uncertainty Milton himself discerned and attempted to stabilize by declaring [in *The Doctrine and Discipline of Divorce*] that this 'was no meer amatorious novel' " (85–86). Yet Sidney seems to anticipate Milton's complaint about the *Arcadia* in his fictive version of Elizabeth's married life—neither a prototype for the realistic novel nor a stale and sentimental romance. In any event, Milton's apparent discomfort with the genres of romance and the novel, like Sidney's qualifications of both, is crucial to what Patterson calls "the emergent poetics of narrative fiction" (86).
59. Tennenhouse (1986) maintains the connections between Sidney's *Arcadia* and "The Letter" in part because "[n]o single writer was as fully conversant with this language of desire as Sidney," for Sidney "not only eroticized the form of romance itself but also domesticated it" (18–19).
60. Godzich and Kittay 179. See also Lamb's (2001) discussion.
61. See Croll 45.

62. McKeon (1987) challenges the long-held critical conjunction of the rise of the novel with the decline of romance narratives. Novelistic prose, McKeon argues, is critical both of romance fictions and of the empiricism that challenged such illusions, and he elsewhere explores a fundamental critical problem, "the inadequacy of our theoretical distinction between 'novel' and 'romance'" (3–6). Patterson adds to McKeon's picture, suggesting that the romance itinerary was at once advanced and derailed by novelistic detail (88).

63. Connell explores the "digression" involving the potion in the trial scene of Book V as a "narrative excursion" that "recapitulates in embryo the main events of the rest of the story. We hear once again of the serious young person overcome by passion, the obstacles to love, the conquest of those obstacles, the elopement of the lovers, and their final reconciliation with their royal parent" (34–35). "The digression implies at the same time," for Connell, "that the larger tale of love in Arcadia has been but one more version of an ancient human story which echoes eternally from generation to generation." "The central motif of the tale," she adds, "has a striking resemblance to *A Defence of Poetry*. The potion with its irresistible effect echoes the 'medicine of cherries' in the *Defence*, serving to disarm those who would resist beauty, itself a 'benign' kind of 'enchantment'" (35–36).

64. Waller (1993) describes the "shattered imaginary" as "a plurality of contradictory narratives" (284).

65. McCoy 26.

Chapter 4   "more like runne-awaies, then Princes"

1. Quoted by Ross 4.

2. See Newcomb (2002) 16; a related discussion is provided by Jameson.

3. Radway explores the genre of romance as a solution to patriarchal contradictions (14), and Waller (1993) describes Wroth's wish to complete a story of her own along some of the lines Radway suggests (279).

4. Brennan 77–78.

5. All references are to *The First Part of the Countess of Montgomery's Urania* edited by Roberts, and henceforth all references will be noted parenthetically in the text.

6. Samuel Wolff's phrase is quoted by McDermott 108.

7. I take this phrasing from Marthe Roberts 24. See also Tennenhouse (1990), who maintains that Sidney's efforts in the *Arcadia* allowed him to command a "certain kind of cultural authority that he lacked in the political world" (210–12).

8. See Lamb (2001) for a discussion of Wroth's romance as a response to the peculiar "dilemma" the Sidney genealogy posed to family members (108).

9. See Hannay (1991) 28. Additional accounts of autobiographical elements in the *Urania* are provided by Hackett (1992); Carrell 92–93; and by Roberts in her edition of *The Poems of Lady Mary Wroth* (30).

10. Laroche comments on this connection from a different angle, comparing Antissia's and Pamphilia's love for Amphilanthus and their sonnets about him, and she claims that their rivalry "establishes a relationship between

reader and writer that is analogous to that between the lover and the beloved, in that these female characters are rivals as readers of the same poet—Sidney—and as lovers of the same man—Amphilanthus, who thus becomes a version of Sidney in this analogy" (276). See also Wynne-Davies (2000) for a linkage of Philip Sidney and William Herbert in Wroth's *Love's Victory* (177).

11. I take this term from Kinney (1995), who explores how Sidney's *New Arcadia* examines romance's "informing values" (37); for a related treatment of Wroth's narrator, see Fendler, who suggests that a "comparison of the different narrative levels reveals a certain kind of hypocrisy among the male characters" (295).

12. See Goldberg's (1986) discussion of absolutism and patriarchy in the Jacobean court (5).

13. Flint describes a similar interest in "competing representations of kinship," arguing that "[e]arly prose fiction, perhaps more than any other genre, merged its interest in family history with its own prolonged form. The encyclopedic nature of the works allowed for the minute examination of family concerns to an extent not usually or easily sustained in other literary genres," "largely because of its flexible incorporation of other discourses such as conduct books, philosophical treatises, and demographic studies" (3–4, 10).

14. Pierre Bourdieu's work ideas are cited by Medick and Sabean, who argue that alongside a focus on intergenerational relations studies of the family should emphasize the interrelationship among siblings. Placing "the centre of interest on reproduction in the dynamics of family life" ignores "the production and reproduction of everyday life," how, e.g., younger sons and sisters regard bonds between fathers and elder sons. "Most considerations of sibling relationships," they note, "centre on marriage strategies of a household, masking the central point that each marriage in a series has repercussions on subsequent marriages" (18–19).

15. Quoted by McClaren 279.

16. Medick and Sabean quote Bourdieu (26 n61).

17. See Flint's discussion of women writers "groping towards realism" as well as his treatment of Aphra Behn's "anti-realist tendencies [that] seem to parallel the fiction's anti-domestic tone" (80–81). These "tendencies" may explain critical discomfort with reading Behn's work in the terms of the novel or the romance. Cf. Armstrong, who makes a case for the way the novel created the realm of domestic space and supported it with women's moral authority: "From the beginning, domestic fiction actively sought to disentangle the language of sexual relations from the language of politics and, in so doing, to introduce a new form of political power" (3). "It is my contention," Armstrong elsewhere elaborates, "that narratives which seemed to be concerned solely with matters of courtship and marriage in fact seized the authority to say what was female, and that they did so in order to contest the reigning notion of kinship relations that attached most power and privilege to certain family lines" (5).

18. Radway 14.

19. I take the term from Bannet 50.
20. See Barber 194.
21. Masten 8.
22. This is from Ben Jonson's tribute to Wroth, cited by Hannay (1991). For related accounts of Wroth's use of the romance form, see Kinney, "Beleeve this"; Lamb, "Topicality and the Interrogation of Wonder"; Krontiris; and Lewalski, who writes: "In the Jacobean era, women appropriated and rewrote available literary and cultural discourses to serve their own needs and interests" (34).
23. See Flint 80. Flint also claims that "prose fiction is engaged in the same institutional process as the family: making relations of power and subordination seem natural and neutral," and he describes prose's "dependence on relational patterns, its strengthening of expressive force through descriptions of subjugation, filiation, and dynastic continuity" (22).
24. I take this formulation from Carrell 80.
25. See Newcomb (2002) 216; and Donovan's discussion of "feminist prosaics" (ix).
26. Quilligan (1989) comments on this episode and more generally upon the *Urania*'s "traffic in women," noting that fathers trade in both sons and daughters, but only brothers trade sisters (269–70).
27. Jameson 134. Lamb (2001) claims that the *Urania* "both engages and evades the logic of genealogical narrative" (108).
28. Thirsk's thinking is cited by Montrose (1991) 32. Thirsk notes that seventeenth-century literary treatments routinely offer comment on and commiseration with their plight. Previously, younger sons, with a sufficient annuity, might go to the university or enter the ministry; a "last refuge" was the Low Countries. Even Edmund Spenser's career follows this trajectory. Thirsk cites John Earle's *Microcosmography* (1633), which proposes that "[t]he planting of settlements in Ireland [was] a means of occupying younger brothers no longer able to enter monasteries" (360, 368). In many ways Philip Sidney lived and died like a younger son, eagerly pursuing a military career, awaiting the fortunes of his uncles and caught up in schemes for colonizing the New World. For related examinations of younger brothers, see Adelman (1985) 91; and Fineman. Austin considers later explorations of the younger brother, and observes that the "most purchased, read, and quoted anti-primogeniture argument of the seventeenth century was John Ap-Robert's 'The Younger Brother, His Apology,' " first published in 1618 and widely circulated when the 1621 *Urania* appears.
29. For an extended investigation of these pressures, see Goldberg (1986).
30. Frye claims: "In every period of history certain ascendant values are accepted by society and are embodied in its serious literature. Usually the process includes some form of kidnapped romance, that is, romance formulas used to reflect certain ascendant religious or social ideals" (28–30, see also 168). See Berger 236; and Jameson (105–07) for applications and modifications of Frye's work.
31. I draw on Tennenhouse's (1990) discussion of these connections (201–03), and on his notion that Sidney "constructed a whole narrative problematic around the question of female inheritance" (203–06). Quilligan (1988)

argues that Sidney acts as a patriarch forcing Elizabeth's hand in marriage, but I would argue that a new form of authority is asserting itself in Sidney's letter to the queen, the moral authority of one's dependents.

For additional explorations of the romance genre's concern with lineage and family boundaries, see Bloch, and Lamb (2001), who explores the *Urania*'s narrative form in terms of a "biopolitics of lineage," so that the "organization of family lines coincides with the appropriation of vernacular literary forms" (75). Likewise, Quilligan (1989) argues: "Given the reliance of romance on family connections, there is a certain generic logic to the fact that, in the history of English literature, the first work of prose fiction published by a woman should be Wroth's romance" (258).

32. As Kinney (1995) writes: "The elaborate knightly devices and 'shows' that play such a large part in the unfinished third book of the romance do not so much celebrate chivalric rituals as place them within a larger design that invites their demystification" (39). Relatively small-scale intrigues in the Old *Arcadia* are replaced, e.g., by civil war in the New *Arcadia*.

33. See Teskey's introduction in Logan and Teskey (3).

34. As Newcomb (2002) comments, "Romance . . . began to drive the work of distinction from the sixteenth century" (8, see also 27).

35. Frye 28.

36. See Naomi J. Miller and Gary Waller's introduction to *Reading Mary Wroth* 1; Waller's (1991) essay "The Sidney Family Romance" 57; and Waller (1993) 269.

37. I take this phrasing from Hanson's discussion of Wroth's Petrarchan poetics (167, 171).

38. McKeon (1987) 317, 339.

39. McKeon (1987) 226. Elsewhere McKeon argues, "[t]he figure of the younger son is central both to the progressive and to the conservative imagination . . . the romance convention of discovered parentage . . . simultaneously confirmed both the nobility of dispossessed younger sons and the ultimate justice of aristocratic culture. But by the seventeenth century, the discrediting of the social convention had severely impaired the efficacy of its literary counterpart" (218–19).

40. Bloch 29. The genre of romance in the sixteenth and seventeenth centuries should be understood not as a proto-novel but as an "alternative [response] to specific social and ideological conditions," "another version of the story of noncorrespondence between virtue and social rank, another mimetic reflection on the mutability of social status." See Newcomb (1996) 230 n10; and (2002) 216. Although Wroth doesn't figure in Newcomb's account, the *Urania* seems to perfectly combine the impulses of Greene and Sidney in challenging so much of what separated elite from popular literature. See also Ballaster 2. Later, Ballaster claims women's amatory fiction in Britain derives from French "feminocentric romances of the 1640, stories with heroines at their center" (44–47).

41. I use Gay's term (13).

42. That these new principles "loosen the rules of primogeniture" is suggested by Tennenhouse (1990) in his discussion of Sidney's final creation of

two families in the *Arcadia*, which each observe the rules of patrilineage (208). See also Montrose (1981) 30.

43. See Goldberg's (1986) discussion of Stuart family images (3–5).

44. McKeon (1987) 133.

45. This literature grows in size and urgency in the course of the seventeenth century. See Thirsk 359–61, quoted by Montrose (1981).

46. See McKeon 1987, especially "The Dialectical Constitution of the Novel," which explores Swift and Defoe's "parables" of younger sons (315–56). Cf. Perry, who says of the eighteenth-century novel: "visible signs of the disinheritance of daughters are everywhere" (113).

47. McKeon (1987) 338–39.

48. See Warner 279–82. Richetti (1969) describes criminal narratives of the seventeenth century similarly: "The criminal's revolt is against social and moral restraints, against any sort of control from an external source. His drive is towards self-determination, primarily and overtly economic, but inescapably spiritual and ideological as well" (52–53). Flint discusses "prodigal fiction" (117 ff.).

49. Perhaps this is also why numerous male characters "prefer the title 'knight' to the title 'prince,' " although Fendler argues that Wroth's narrator "equates" this preference "with the evasion of duties they might have to their countries and families" (295). This is certainly not the case with younger brothers in the *Urania*. Often left only with "the empty pack of disinheritance," younger brothers are not unlike the women in Wroth's romance who love them or the early women writers who so frequently tell their stories. See Gilbert, who describes "the conundrum of the empty pack which until recently has confronted every woman writer. . . . the riddle of daughterhood, a figurative empty pack with which, it has seemed to many women artists, every powerful literary mother as well as every literal mother presents her daughter" (257–58).

50. Kolodny 5. Donovan describes the early novel's intense interest in economic realities (4).

51. See Backscheider and Richetti's introduction (xiv).

52. In their introduction, Backscheider and Richetti emphasize these features as attributes of later women's fiction (ix–x).

53. Quilligan (1989) emphasizes the brother's power over the sister: "The women silently suffer not only their wayward lovers," Quilligan argues, "but their brothers' wayward action" (267). Quilligan notes too: "[t]he sister's story is only tellable as the brother's tale; the only social identity possible for a woman at this time is established through the link with the brother" (280).

54. Cf. Lewalski's claim that Wroth "used her heritage transgressively to replace heroes with heroines at the center of several major genres employed by the male Sidney authors, transforming their values and gender politics and exploring the poetics and situations of women writers" (244).

55. Montrose (1981) claims, e.g., that Shakespeare's plays "explore the difficulty or impossibility of establishing or authenticating a self in a rigorously hierarchical and patriarchal society, a society in which full social identity

tends to be limited to propertied adult males who are the heads of households" (35).

56. Shaver 75 n11.

57. Lamb (2001) discusses the narrative shape of Wroth's debts to Sidney (112–13). See also Quilligan (1990); and MacArthur.

58. See Waller's (1991) essays "The Sidney Family Romance" (38) and "Mother/Son," where he writes: "The intriguing contradictions in this corner of the family romance require more than pieties about Wroth's aunt's (or father's) 'influence' or 'encouragement' since Wroth was '[d]ominated by male authority figures—first her father, then (however minimally) her husband, and (over probably the course of twenty years) her cousin-lover' " (409).

59. Segre 25.

60. For connections to Montemayor, see Hackett (2000); and Laroche 268. Cavanagh, who claims that Wroth, with access to her family's library and learning, was well versed in the scientific and philosophical controversies of her day (5–6). See also McClaren 278.

61. See Teskey's introduction to Logan and Teskey (4).

62. Newcomb (2002) 214; and Warner 280–84.

63. Goldberg (1986) 20.

64. Lamb (2001) 107.

65. Lamb (2001) 108, drawing on Bloch 96.

66. Bloch 28.

67. I take the term from Radway 162–62.

68. Jameson 104.

69. Jameson 105. See too Bloch, who argues: "Descent was a less potent force of family cohesion than affiliation with living relatives" (66). Quint similarly describes romance as a horizontal world of action. For an investigation of the differences between these two kinds of plots, see Spacks 114–46. See also Donovan's discussion of the epic as canonizing the world of the fathers (3–4).

70. Jameson 118–19.

71. Additional biographical details are offered by Quilligan (1990) 307; and Lamb (2001) 110–11.

72. Cf. Goldberg (1986) 7.

73. Quoted by Goldberg (1986) 8.

74. See Richetti 18.

75. See Flint 86.

76. Flint 83.

77. Austin argues: "[d]uring and immediately after the Civil War, English primogeniture . . . came under attack by both religious and political radicals," and describes how religious reformers were "spurred on by nearly a dozen key biblical narratives featuring younger sons supplanting their older brothers" (13; see also 22 n3).

78. McKeon (1987) 133; see also 19.

79. Quoted by McKeon (2000) 160; taken from Marthe Roberts 21. Roberts goes on to describe the novel as related to "a certain type of elementary storytelling, half-way between literature and psychology." "The dual psychological and literary character of this rediscovered myth," she adds, "the originality of

its structure, the peculiarity of its content and the pathological nature of its revival could not have been more aptly epitomised than in Freud's now classical definition: 'the family romance of the neurotic' " (22).

80. Carrell quotes Arthur Wilson's *The History of Great Britain* (1653) 92.

Epilogue   "Curious Networks" and Lost Sons

1. All references to *Muiopotmos: Or the Fate of the Butterflie* are to *The Yale Edition of the Shorter Poems of Edmund Spenser*.

2. I am obviously indebted to Lemmi, who contends that Spenser's poem is an allegory of the life and death of Sidney, "much like *Astrophel*": "In both we see a gracious and gifted being who leaves behind him his youthful sports and the languishing eyes of many fair ones to go adventuring. In both he plunges recklessly into the dangerous pleasure he seeks. In both he is suddenly slain" (732). I disagree, though, with many of the biographical connections Lemmi draws with other personages, e.g., between Astery and Stella; and I take issue with Lemmi's claim that Spenser regarded Sidney "as having been foully done to death, a noble victim to the hatred of the enemies of the Queen" (732). A summary of earlier critical approaches to *Muiopotmos* is offered by Court.

3. See Andrew D. Weiner 213–14.

4. Hager (1990) notes: "although he is capable . . . of perhaps the first and certainly some of the most spectacular flights of Golden poetry in England . . . Sidney remains at heart—paradoxically—a deflator of the strained ideals such poetry expresses, or, shall we say, a critic of the dangers to identity and rank that inevitable yearning for those ideals entails" (488).

5. I borrow the term from Kassanoff 61.

6. Greville, *The Life of the Renowned Sir Philip Sidney* (1652) 19–20.

7. Duncan-Jones (1991) 3.

8. See Anderson 104, cited by Bond.

9. Lipking argues: "the development of a great many poets follows a consistent internal logic" (viii). Yet in order for this to be true, poets need first to think of themselves as poets, a "development" we first encounter in Spenser's work, not in Sidney's. Pask explains what had to happen to Sidney's story in order to recuperate him as an author, and suggests, "the transition to the status of poet occurred along with the derogation of [Sidney's] earlier cultural authority as an aristocratic hero" (5).

10. *DP* 17. See also Hager (1990) 1. Hager (1990) argues for an image of Sidney as "a critic of human aspiration, a didactic poet working by indirection and irony even in his own behavior, a master of the overview that his exemplary image would deny him" (2) and offers a shrewd analysis of Sidney's "corrective irony" (10).

11. Steinberg (1994) 192. Steinberg also comments that "Spenser's honesty, his refusal to accept the Sidney myth, reflects his state of mind in the 1590s, his disillusionment with the court and with the Elizabethan world. Because the poet could try to improve that world, Spenser had remained true to his calling; but Sidney, he regretted, had not" (194). See also Bond's notes to

Muiopotmos in the *Yale Edition* of Spenser's poems: "The change in the vowel [from Astrophil to Astrophel] quietly substitutes for the aspiration implicit in the original name ('star-lover') the suggestion of a star fallen" (569).

12. Steinberg (1994) 187.
13. Tillich 50. See also Gibbs 22.
14. See Lipking ix.
15. Don Cameron Allen's comment is cited by Andrew D. Weiner 206.
16. McNeir 70–71.
17. Lanham 329.
18. Cf. Anderson 91, 94.
19. I take the phrase from Mason 15–19.
20. See the edition of Sidney's prose works prepared by Feuillerat (3: 140); cited by Lanham 330.
21. Jameson 87.
22. See Blisset.
23. Blisset 89.
24. Brinkley 674.
25. See Brinkley 674.
26. Brinkley 674.
27. As Anderson puts it: "The irony—or the web—in which Clarion is trapped takes shape in the myth of Astery" (97). Anderson also notes, "Clarion's present beauty is of mixed birth, formed by Love (Venus), but also by Hatred (envy, slander, fear, jealousy). There is an ugly malice in the world to which the unwitting Clarion takes his flight, but there is an ugliness and weakness in Clarion's background as well" (100).
28. Quoted by Dundas 129.
29. Pask describes the "extent of this misrecognition of early modern 'friendship' " specifically in the case of Spenser and Sidney, raising an important point about the "construction of a literary 'kinship' between Sidney and Spenser—a process that began with Spenser himself—[and] transformed Renaissance patron-client relations into narratives of interpoetic influence and quasi-aristocratic national 'patrimony' " (83–84).
30. All references to *The Faerie Queene* are to the edition prepared by A.C. Hamilton (NY: Longman, 1977, 1987, 2001).
31. As Steinberg (1994) claims, "Astrophel is like a Redcross Knight who dies in his clash with Orgoglio" (190, 191).
32. Berger (1988) 28.
33. See Anderson 97–100.
34. Nancy K. Miller 272–73, 275.
35. See Greenblatt (1973) 271; cited by McCoy 29.
36. See Helgerson (1983) 72. Helgerson describes the scene at Acidale a bit differently, viewing the spectacle as Spenser's way of pitting amateur poets against poets laureate: "the two poets, or the two ideas of poetry, no longer cohere. The private poet rebels against his public duty; the public poet can find no use for his private inspiration" (83).
37. John Milton, *Complete Poems and Major Prose*, ed. Merritt Y. Hughes (NY: Macmillan, 1985).

BIBLIOGRAPHY

Primary Works of the Sidney Family

Herbert, Mary Sidney. *The Collected Works of Mary Sidney Herbert. Countess of Pembroke. Volume 1. Poems, Translations, and Correspondence.* Margaret P. Hannay, Noel J. Kinnamon, and Michael G. Brennan, editors. Oxford: Clarendon Press, 1998.

The Psalms of Sir Philip Sidney and The Countess of Pembroke. J. C. A. Rathmell, editor. NY: New York University Press, 1963.

Sidney, Sir Philip. *A Defence of Poetry.* J.A. Van Dorsten, editor. NY: Oxford University Press, 1966, 1997.

———. *A Defence of the Earl of Leicester* (1584). Autograph version. The Pierpont Morgan Library. MA 1475.

———. *The Countess of Pembroke's Arcadia (The Old Arcadia).* Jean Robertson, editor. Oxford: Clarendon Press, 1973.

———. *Miscellaneous Prose of Sir Philip Sidney.* Katherine Duncan-Jones and Jan Van Dorsten, editors. Oxford: Clarendon Press, 1973.

———. *Miscellaneous Works of Sir Philip Sidney.* William Grady, editor. Boston: T.O.H.P. Burnham, 1860.

———. *The Old Arcadia.* Katherine Duncan-Jones, editor. NY: Oxford University Press, 1985, 1994.

———. *The Poems of Sir Philip Sidney.* William A. Ringler, Jr., editor. Oxford: Clarendon Press, 1962.

———. *The Prose Works of Sir Philip Sidney.* Albert Feuillerat, editor. London: Cambridge University Press, 1962–69.

———. *Sir Philip Sidney: A Critical Edition of the Major Works.* Katherine Duncan-Jones, editor. NY: Oxford University Press, 1989.

Sidney, Robert. *The Poems of Robert Sidney, Ed. from the Poet's Autograph Notebook with an Introduction and Commentary.* P.J. Croft, editor. Oxford: Clarendon Press, 1984.

———. "The Poems of Sir Robert Sidney." Katherine Duncan-Jones, editor. *English* 30, 136 (1981): 3–71.

Wroth, Lady Mary. *The First Part of the Countess of Montgomery's Urania*, MRTS 140. Josephine A. Roberts, editor. Binghamton, NY: Medieval and Renaissance Texts and Studies, 1995.

———. *The Poems of Lady Mary Wroth.* Josephine A. Roberts, editor. Baton Rouge: Louisiana State University Press, 1983.

Secondary Works

Adelman, Janet. "Male Bonding in Shakespeare's Comedies." *Shakespeare's "Rough Magic": Renaissance Essays in Honor of C.L. Barber.* Peter Erickson and Coppelia Kahn, editors. Newark, DE: University of Delaware Press, 1985. 73–103.

———. *Suffocating Mothers: Fantasies of Maternal Origin in Shakespeare's Plays, Hamlet to The Tempest.* NY: Routledge, 1992.

Adolph, Robert. *The Rise of Modern Prose Style.* Cambridge, MA: The MIT Press, 1968.

Altman, Joel. *The Tudor Play of Mind: Rhetorical Inquiry and the Development of Elizabethan Drama.* Berkeley: University of California Press, 1978.

Anderson, Judith. " 'Nat worth an boterflye': *Muiopotmos* and *The Nun's Priest's Tale.*" *Journal of Medieval and Renaissance Studies* 1 (1971): 89–106.

Archer, John Michael. *Sovereignty and Intelligence: Spying and Court Culture in the English Renaissance.* Stanford, CA: Stanford University Press, 1993.

Armstrong, Nancy. *Desire and Domestic Fiction: A Political History of the Novel.* NY: Oxford University Press, 1987.

Austin, Michael. "Aphra Behn, Mary Pix, and the Sexual Politics of Primogeniture." *Restoration and Eighteenth Century Theatre Research* 16, 1 (2001): 13–23.

Backscheider, Paula and John J. Richetti, editors. *Popular Fiction by Women 1660–1730. An Anthology.* Oxford and New York: Oxford University Press, 1996.

Bakhtin, M.M. *The Dialogic Imagination.* Michael Holquist, editor. Austin: University of Texas Press, 1981.

Ballaster, Ros. *Seductive Forms: Women's Amatory Fiction from 1684–1740.* NY: Oxford University Press, 1992.

Bannet, Eve Tavor. *The Domestic Revolution: Enlightenment Feminisms and the Novel.* Baltimore: Johns Hopkins University Press, 2000.

Barber, C.L. "The Family in Shakespeare's Development: Tragedy and Sacredness." *Representing Shakespeare: New Psychoanalytic Essays.* Murray M. Schwartz and Coppelia Kahn, editors. Baltimore: Johns Hopkins University Press, 1980. 188–202.

Barkan, Leonard. *The Gods Made Flesh: Metamorphosis and the Pursuit of Paganism.* New Haven: Yale University Press, 1986.

Bateson, Gregory. "Toward a Theory of Schizophrenia." *Steps to an Ecology of Mind: Collected Essays in Anthropology, Psychiatry, Evolution, and Epistemology.* San Francisco: Chandler Publishing Company, 1972. 201–27.

Bellamy, Elizabeth J. "Psychoanalysis and the Subject in/of/for the Renaissance." *Bucknell Review* 35, 2 (1992): 19–33.

Belsey, Catherine. *Shakespeare and the Loss of Eden: The Construction of Family Values in Early Modern Culture.* New Brunswick, NJ: Rutgers University Press, 1999.

Berger, Jr., Harry. " 'Kidnapped Romance': Discourse in *The Faerie Queene.*" *Unfolded Tales: Essays on Renaissance Romance.* George M. Logan and Gordon Teskey, editors. Ithaca: Cornell University Press, 1989. 208–56.

——. *Revisionary Play: Studies in the Spenserian Dynamics.* Berkeley: University of California Press, 1988.

Berry, Edward I. *The Making of Sir Philip Sidney.* Toronto: University of Toronto Press, 1998.

Blisset, William. "Florimell and Marinell." *Studies in English Literature* 5 (1965): 87–104.

Bloch, R. Howard. *Etymologies and Genealogies: A Literary Anthropology of the French Middle Ages.* Chicago: University of Chicago Press, 1983.

Blumenberg, Hans. *Work on Myth.* Robert M. Wallace, translator. Cambridge, MA: The MIT Press, 1985.

Bond, Ronald B. "*Invidia* and the Allegory of Spenser's *Muiopotmos.*" *English Studies in Canada* 2 (1976): 144–55.

Bono, Barbara J. "'The Chief Knot of All the Discourse': The Maternal Subtext Tying Sidney's *Arcadia* to Shakespeare's *King Lear.*" *Gloriana's Face: Women, Public and Private, in the English Renaissance.* S.P. Cerasano and Marion Wynne-Davies, editors. Detroit: Wayne State University Press, 1992. 105–27.

Boose, Lynda E., "The Family in Shakespeare Studies; or—Studies in the Family of Shakespeareans; or—The Politics of Politics." *Renaissance Quarterly* 40, 4 (1987): 707–42.

Brennan, Michael G. "Creating female authorship in the early seventeenth century: Ben Jonson and Mary Wroth." *Women's Writing and the Circulation of Ideas: Manuscript Publication in England, 1550–1800.* George L. Justice and Nathan Tinker, editors. NY: Cambridge University Press, 2000. 73–93.

Brinkley, Robert A. "Spenser's *Muiopotmos* and the Politics of Metamorphosis." *English Literary History* 48 (1981): 668–76.

Buxton, John. "The Mourning for Sidney." *Renaissance Studies* 3 (1989): 46–56.

——. *Sir Philip Sidney and the English Renaissance.* NY: St. Martin's Press, 1964.

Campbell, Lily B. *Divine Poetry and Drama in Sixteenth-Century England.* Berkeley: University of California Press, 1959.

Cantar, Brenda. "Charmed Circles of Enchantment: Pre-oedipal Fantasies in Sir Philip Sidney's *Arcadia.*" *Sidney Newsletter and Journal* 12, 1 (1992): 3–20.

Carey, John. "Elizabethan Prose." *English Poetry and Prose 1540–1674.* Christopher Ricks, editor. *Sphere History of Literature in the English Language.* Volume 2. London: Sphere Books, 1970. 361–89.

Carrell, Jennifer Lee. "A Pack of Lies in a Looking Glass: Lady Mary Wroth's *Urania* and the Magic Mirror of Romance." *Studies in English Literature* 34 (1994): 79–107.

Cavanagh, Sheila T. *Cherished Torment: The Emotional Geography of Lady Mary Wroth's Urania.* Pittsburgh: Duquesne University Press, 2001.

Cavell, Stanley. "The Avoidance of Love: A Reading of *King Lear.*" *Disowning Knowledge in Six Plays of Shakespeare.* NY: Cambridge University Press, 1987. 39–124.

Coch, Christine. "'Mother of my Contreye': Elizabeth I and Tudor Constructions of Motherhood." *English Literary Renaissance* 26, 3 (1996): 423–50.

Connell, Dorothy. *Sir Philip Sidney: The Maker's Mind.* Oxford: Clarendon Press, 1977.

Corthell, Ronald J. "The Subject of Nonfictional Prose: The Renaissance." *Prose Studies* 11, 2 (1988): 3–9.

Court, Franklin E. "The Theme and Structure of Spenser's *Muiopotmos.*" *Studies in English Literature* 10 (1970): 1–15.

Cressy, David. "Kinship and Kin Intervention in Early Modern England." *Past and Present* 113 (1986): 38–69.

Crewe, Jonathan. *Hidden Designs: The Critical Profession and Renaissance Literature.* NY: Methuen, 1986.

Croll, Morris. *Style, Rhetoric, and Rhythm: Essays by Morris Croll.* J. Max Patrick et al., editors. Princeton: Princeton University Press, 1966.

Danby, John. *Elizabethan and Jacobean Poets.* London: Faber and Faber, 1965.

Davis, Natalie Zemon, "Ghosts, Kin, and Progeny: Some Features of Family Life in Early Modern France." *Daedulus* 106 (1977): 87–114.

Davis, Walter. *A Map of Arcadia: Sidney's Romance in Its Tradition.* Richard A. Lanham. *The Old Arcadia.* Yale Studies in English. Volume 158. New Haven: Yale University Press, 1965.

Diefendorf, Barbara B. "Family Culture, Renaissance Culture." *Renaissance Quarterly* 40, 4 (1987): 661–81.

Dipple, Elizabeth. "Metamorphosis in Sidney's *Arcadias.*" *Philological Quarterly* 50 (1971): 47–62.

Dolan, Frances E. *Whores of Babylon: Catholicism, Gender, and Seventeenth-Century Print Culture.* Ithaca: Cornell University Press, 1999.

Donovan, Josephine. *Women and the Rise of the Novel, 1405–1726.* NY: St. Martin's Press, 2000.

Duncan-Jones, Katherine. "Sir Philip Sidney's Toys." *Proceedings of the British Academy* 66 (1980): 161–78.

———. *Sir Philip Sidney: Courtier Poet.* New Haven: Yale University Press, 1991.

Dundas, Judith. "'The Heaven's Ornament': Spenser's Tribute to Sidney." *Etudes Anglaises* 42, 2 (1989): 129–39.

Elizabethan and Jacobean Prose 1550–1620. Volume 1 of *The Pelican Book of English Prose.* Introduction by Kenneth Muir. Harmondsworth: Pelican, 1956.

Esler, Anthony. *The Aspiring Mind of the Elizabethan Younger Generation.* Durham, NC: Duke University Press, 1966.

Falco, Raphael. *Conceived Presences: Literary Genealogy in Renaissance England.* Amherst: University of Massachusetts Press, 1994.

Feinstein, Howard M. "A Singular Life: Twinship in the Psychology of William and Henry James." *Blood Brothers: Siblings as Writers.* Norman Kiell, editor. NY: International Universities Press, 1983. 301–28.

Fendler, Susanne. "Questioning the Knight's Quest: The Narrator as Judge in Two Imitations of Sidney's *Arcadia.*" *Narrative Strategies in Early English Fiction.* Wolfgang Gortschacher and Holger Klein, editors. Lewiston, NY: Edwin Mellen Press, 1995. 289–305.

Ferguson, Margaret W. *Trials of Desire: Renaissance Defences of Poetry.* New Haven: Yale University Press, 1983.

Ferry, Anne. *The "Inward" Language: Sonnets of Wyatt, Sidney, Shakespeare, Donne.* Chicago: University of Chicago Press, 1983.

Fineman, Joel. "Fratricide and Cuckoldry: Shakespeare's Doubles." *Representing Shakespeare: New Psychoanalytic Essays*. Murray M. Schwartz and Coppelia Kahn, editors. Baltimore: Johns Hopkins University Press, 1980. 70–109.

Fisken, Beth Wayne. "'To the Angel Spirit...': Mary Sidney's Entry into the 'World of Words.'" *The Renaissance Englishwoman in Print: Counterbalancing the Canon*. Anne M. Haselkorn and Betty S. Travitsky, editors. Amherst: University of Massachusetts Press, 1990. 263–75.

Flint, Christopher. *Family Fictions: Narrative and Domestic Relations in Britain, 1688–1798*. Stanford: Stanford University Press, 1998.

Fowler, Elizabeth and Roland Greene, editors. *The Project of Prose in Early Modern Europe and the New World*. NY: Cambridge University Press, 1997.

Fraser, Antonia. *Mary Queen of Scots*. NY: Dell, 1969.

Freccero, John. "The Fig Tree and the Laurel: Petrarch's Poetics." *Literary Theory/Renaissance Texts*. Patricia Parker and David Quint, editors. Baltimore: Johns Hopkins University Press, 1986. 20–32.

Freud, Sigmund. "Inhibition, Symptoms and Anxiety." (1926) *The Standard Edition of the Complete Psychological Works of Sigmund Freud*. Volume 20. James Strachey et al., editors. London: Hogarth Press, 1953–74.

Frye, Northrop. *The Secular Scripture: A Study of the Structure of Romance*. Cambridge, MA: Harvard University Press, 1976.

Fumerton, Patricia. *Cultural Aesthetics: Renaissance Literature and the Practice of Social Ornament*. Chicago: University of Chicago Press, 1991.

Gay, Peter. *The Bourgeois Experience. Victoria to Freud. Volume 1. The Education of the Senses*. NY: Oxford University Press, 1984.

Gibbs, Lee W. "Myth and the Mystery of the Future." *Myth and the Crisis of Historical Consciousness*. Lee W. Gibbs and Taylor Stevenson, editors. Missoula, MT: Scholar's Press, 1975. 19–33.

Gilbert, Sandra M. "Life's Empty Pack: Notes toward a Literary Daughteronomy." *Daughters and Fathers*. Lynda E. Boose and Betty S. Flowers, editors. Baltimore: Johns Hopkins University Press, 1989. 355–84.

Godzich, Wlad and Jeffrey Kittay. *The Emergence of Prose: An Essay in Prosaics*. Minneapolis: University of Minnesota Press, 1987.

Goldberg, Jonathan. "The Countess of Pembroke's Literal Translation." *Subject and Object in Renaissance Culture*. Margreta De Grazia, Maureen Quilligan, and Peter Stallybrass, editors. NY: Cambridge University Press, 1996. 321–36.

——. "Fatherly Authority: The Politics of Stuart Family Images." *Rewriting the Renaissance: The Discourses of Sexual Difference in Early Modern Europe*. Margaret W. Ferguson, Maureen Quilligan, and Nancy J. Vickers, editors. Chicago: University of Chicago Press, 1986. 3–32.

Goody, Jack. *The Development of the Family and Marriage in Europe*. Cambridge, UK: Cambridge University Press, 1983.

Greenblatt, Stephen. "Psychoanalysis and Renaissance Culture." *Literary Theory/Renaissance Texts*. Patricia Parker and David Quint, editors. Baltimore: Johns Hopkins University Press, 1985. 210–24.

——. *Renaissance Self-Fashioning from More to Shakespeare*. Chicago: University of Chicago Press, 1980.

Greenblatt, Stephen. "Sidney's *Arcadia* and the Mixed Mode." *Studies in Philology* 70 (1973): 269–78.

Greene, Roland. "Fictions of Immanence, Fictions of Embassy." *The Project of Prose in Early Modern Europe and the New World*. Elizabeth Fowler and Roland Greene, editors. NY: Cambridge University Press, 1997. 176–202.

——. *Post-Petrarchism: Origins and Innovations of the Western Lyric Sequence*. Princeton: Princeton University Press, 1991.

——. "Sir Philip Sidney's *Psalms*, The Sixteenth-Century Psalter, and the Nature of Lyric." *Studies in English Literature* 30, 1 (1990): 19–40.

Greville, Fulke. *The Life of the Renowned Sir Philip Sidney* (1652). Delmar, NY: Scholar's Facsimiles and Reprints, 1984.

——. *The Prose Works of Fulke Greville, Lord Brooke*. John Gouws, editor. Oxford: Clarendon Press, 1986.

Guillory, John. *Poetic Authority: Spenser, Milton, and Literary History*. NY: Columbia University Press, 1983.

Hackett, Helen. "Courtly Writing by Women." *Women and Literature in Britain 1500–1700*. Helen Wilcox, editor. NY: Cambridge University Press, 1998. 169–89.

——. *Women and Romance Fiction in the English Renaissance*. NY: Cambridge University Press, 2000.

——. " 'Yet Tell Me Some Such Fiction': Lady Mary Wroth's *Urania* and the 'Femininity' of Romance." *Women, Texts, and Histories 1575–1760*. Clare Brant and Diane Purkiss, editors. London and NY: Routledge, 1992. 39–68.

Hager, Alan. *Dazzling Images: The Masks of Sir Philip Sidney*. Cranbury, NJ: Associated University Presses, 1991.

——. "Rhomboid Logic: Anti-Idealism and a Cure for Recusancy in Sidney's *Lady of May*." *English Literary History* 57 (1990): 485–502.

Hamilton, A.C. "Elizabethan Prose Fiction and Some Trends in Recent Criticism." *Renaissance Quarterly* 37, 1 (1984): 21–33.

Hannay, Margaret P. " 'Bearing the livery of your name': The Countess of Pembroke's Agency in Print and Scribal Publication." *The Sidney Journal* 18, 1 (2000): 7–42.

——. " 'House-confined maids': The Presentation of Women's Role in the *Psalmes* of The Countess of Pembroke." *English Literary Renaissance* 24, 1 (1994): 44–71.

——. " 'So May I With the *Psalmist* Truly Say': Early Modern Englishwomen's Psalm Discourse." *Write or Be Written. Early modern women poets and cultural constraints*. Barbara Smith and Ursula Appelt. Burlington, VT: Ashgate, 2001. 105–34.

——. " 'This Moses and This Miriam': The Countess of Pembroke's Role in the Legend of Sir Philip Sidney." *Sir Philip Sidney's Achievements*. M.J.B. Allen et al., editors. NY: AMS Press, 1990. 217–26.

——. " 'Your vertuous and learned Aunt': The Countess of Pembroke as a Mentor to Mary Wroth." *Reading Mary Wroth: Representing Alternatives in Early Modern England*. Naomi J. Miller and Gary Waller, editors. Knoxville: University of Tennessee Press, 1991. 15–34.

Hanson, Elizabeth. "Boredom and Whoredom: Reading Renaissance Women's Sonnet Sequences." *The Yale Journal of Criticism* 10, 1 (1997): 165–91.

Hawkins, Peter S. "From Mythography to Myth-Making: Spenser and the Magna Mater Cycle." *Sixteenth-Century Journal* 12, 3 (1981): 51–64.

Hay, Millicent V. *The Life of Robert Sidney.* Washington, DC: The Folger Shakespeare Library, 1984.

Helgerson, Richard. *The Elizabethan Prodigals.* Berkeley: University of California Press, 1976.

——. *Self-Crowned Laureates: Spenser, Jonson, Milton, and the Literary System.* Berkeley: University of California Press, 1983.

Hill, Christopher. "Sex, Marriage, and the Family in England." *Economic History Review.* 2nd series. 31 (1978): 450–63.

Hopkins, John T. "'Such a Twin Likeness There Was in the Pair': An Investigation in the Painting of the Cholmondeley Sisters." *Transactions of the Historical Society of Lancashire and Cheshire* 141 (1991): 1–37.

Hulse, Clark. "Stella's Wit: Penelope Rich as Reader of Sidney's Sonnets. *Rewriting the Renaissance: The Discourse of Sexual Difference in Early Modern Europe.* Margaret W. Ferguson, Maureen Quilligan, and Nancy J. Vickers, editors. Chicago: University of Chicago Press, 1986. 272–86.

Imbrie, Ann E. "Defining Nonfiction Genres." *Renaissance Genres: Essays on Theory, History, and Interpretation.* Barbara Kiefer Lewalski, editor. Cambridge, MA: Harvard University Press, 1986. 45–69.

James. J.B. "The Other Sidney." *History Today* 15 (1965): 183–90.

Jameson, Fredric. "Magical Narratives: On the Dialectical Use of Genre Criticism." *The Political Unconscious: Narrative as a Socially Symbolic Act.* Ithaca: Cornell University Press, 1981. 103–50.

Kalstone, David. *Sidney's Poetry: Contexts and Interpretations.* Cambridge, MA: Harvard University Press, 1965.

Kassanoff, Jennie A. "Extinction, Taxidermy, Tableaux Vivants: Staging Race and Class in *The House of Mirth*." *Publications of the Modern Language Association of America* 115, 1 (2000): 60–74.

Kay, Dennis. "'She was a Queen, and Therefore Beautiful': Sidney, His Mother, and Queen Elizabeth." *Research in English Studies.* New series 43, 169 (1992): 18–39.

Kelliher, Hilton and Katherine Duncan-Jones, "A Manuscript of Poems by Robert Sidney: Some Early Impressions." *British Library Journal* 1 (1975): 107–44.

Kinney, Clare R. "'Beleeve this butt a fiction': Female Authorship, Narrative Undoing and the Limits of Romance in *The Second Part of the Countess of Montgomery's Urania*" (forthcoming in *Spenser Studies*).

——. "Chivalry Unmasked: Courtly Spectacle and the Abuses of Romance in Sidney's *New Arcadia.*" *Studies in English Literature* 35 (1995): 35–52.

——. "Hybrid Text or Hypertext?: Mary Wroth's Recombinant Poetics of Genre." (unpublished essay)

Klein, Lisa M. *The Exemplary Sidney and the Elizabethan Sonneteer.* Newark, DE: University of Delaware Press, 1998.

Klein, Melanie. *The Psychoanalysis of Children.* Translated Alix Strachey. Revised edition. NY: Delacorte Press, 1975.

Kolodny, Annette. "Dancing Through the Minefield: Some Observations on the Theory, Practice, and Politics of a Feminist Literary Criticism." *Feminist Studies* 6, 1 (1980): 1–25.

Krier, Theresa. "Generations of Blazons: Psychoanalysis and the Song of Songs in the *Amoretti.*" *Texas Studies in Literature and Language* 40, 3 (1998): 293–327.

Krontiris, Tina. *Oppositional Voices: Women as Writers and Translators of Literature in the English Renaissance.* London and NY: Routledge, 1992, 1997.

Lamb, Mary Ellen. "Apologizing for Pleasure in Sidney's *Apology for Poetry*: The Nurse of Abuse Meets the Tudor Grammar School." *Criticism* 36, 4 (1994): 499–519.

———. "The Biopolitics of Romance in Mary Wroth's *The Countess of Montgomery's Urania.*" *English Literary Renaissance* 31 (2001): 121–30.

———. "Exhibiting Class and Displaying the Body in Sidney's *Countess of Pembroke's Arcadia.*" *Studies in English Literature* 37, 1 (1997): 55–72.

———. *Gender and Authorship in the Sidney Circle.* Madison, WI: University of Wisconsin Press, 1990.

———. "Topicality and the Interrogation of Wonder in Mary Wroth's *The Second Part of The Countess of Montgomery's Urania*" (forthcoming in *Historicisms.* Anne Lake Prescott, ed.).

Lanham, Richard. "Sidney: The Ornament of His Age." *Southern Review* 2 (1967): 319–40.

Laroche, Rebecca. "Pamphilia Across a Crowded Room: Mary Wroth's Entry into Literary History." *Genre* 30 (1997): 267–88.

Laslett, Peter. "The Bastardy Prone Sub-Society." *Bastardy and Its Comparative History.* Peter Laslett, Karla Oosterveen, and Richard M. Smith, editors. Cambridge, MA: Harvard University Press, 1980. 217–46.

Laslett, Peter and Richard Wall, editors. *Household and Family in Past Time.* Cambridge, UK: Cambridge University Press, 1972.

Lemmi, C.W. "The Allegorical Meaning of Spenser's *Muiopotmos.*" *Publications of the Modern Language Association of America* 45 (1930): 732–48.

Levy, F.J. "Philip Sidney Reconsidered." *Sidney in Retrospect: Selections from English Literary Renaissance.* Arthur F. Kinney, editor. Amherst: University of Massachusetts Press, 1988. 3–14.

Lewalski, Barbara Kiefer. *Writing Women in Jacobean England.* Cambridge, MA: Harvard University Press. 1993.

Lipking, Lawrence. *The Life of the Poet: Beginning and Ending Poetic Careers.* Chicago: University of Chicago Press, 1981.

Logan, George M. and Gordon Teskey, editors. *Unfolded Tales: Essays on Renaissance Romance.* Ithaca: Cornell University Press, 1989.

Loewenstein, Joseph. "Sidney's Truant Pen." *Modern Language Quarterly* 46 (1985): 128–42.

MacArthur, Janet. " 'A Sydney, Though Un-named': Lady Mary Wroth and her Poetical Progenitors." *English Studies in Canada* 25 (1989): 12–30.

Macchiavelli, Niccolo. *The Prince, and The Discourses*; with an introduction by Max Lerner. NY: Modern Library, 1950.

Marotti, Arthur. " 'Love is not Love': Elizabethan Sonnet Sequences and the Social Order." *English Literary History* 49 (1982): 396–428.

Martin, Christopher. " 'Misdoubting His Estate': Dynastic Anxiety in Sidney's *Arcadia*." *English Literary Renaissance* 18 (1988): 369–88.

Maslen, R.W. *Elizabethan Fictions: Espionage, Counter-Espionage, and the Duplicity of Fiction in Elizabethan Prose Narratives*. Oxford: Clarendon University Press, 1997.

Mason, Herbert. "Myth as an 'Ambush of Reality.' " *Myth, Symbol, and Reality*. Alan M. Olson, editor. Notre Dame, IN: University of Notre Dame Press, 1980. 15–19.

Masten, Jeff. " 'Shall I Turn Blabb?': Circulation, Gender, and Subjectivity in Mary Wroth's Sonnets." *Reading Mary Wroth: Representing Alternatives in Early Modern England*. Naomi J. Miller and Gary Waller, editors. Knoxville: University of Tennessee Press, 1991. 67–87.

May, Steven W. *The Elizabethan Courtier Poets: The Poems and Their Contexts*. Columbia, MO: University of Missouri Press, 1991.

Mazzaro, Jerome. *Transformations in the Renaissance English Lyric*. Ithaca: Cornell University Press, 1970.

Mazzola, Elizabeth. "Marrying Medusa: Spenser's *Epithalamion* and Renaissance Reconstructions of Female Privacy." *Genre* 25, 2–3 (1992): 193–210.

McCoy, Richard. *Sir Philip Sidney: Rebellion in Arcadia*. New Brunswick, NJ: Rutgers University Press, 1979.

McDermott, Hubert. *Novel and Romance: The Odyssey to Tom Jones*. Totowa, NJ: Barnes and Noble Books, 1989.

McKeon, Michael. *The Origins of the English Novel, 1600–1740*. Baltimore: Johns Hopkins University Press, 1987.

McKeon, Michael, editor. *Theory of the Novel: A Historical Approach*. Baltimore: Johns Hopkins University Press, 2000.

McLaren, Margaret Anne. "An Unknown Continent: Lady Mary Wroth's Forgotten Pastoral Drama, 'Love's Victorie.' " *The Renaissance Englishwoman in Print: Counterbalancing the Canon*. Anne M. Haselkorn and Betty S. Travitsky, editors. Amherst: University of Massachusetts Press, 1990. 276–94.

McNeir, Waldo. "Spenser and the Myth of Queen Elizabeth." *Spenser: Classical, Medieval, Renaissance, and Modern*. Cleveland: Cleveland State University Press, 1977. 57–71.

Medick, Hans and David Sabean. *Interest and Emotion: Essays on the Study of Family and Kinship*. NY: Cambridge University Press, 1984.

Miller, Nancy K. "Arachnologies: The Woman, the Text, and the Critic." *The Poetics of Gender*. Nancy K. Miller, editor. NY: Columbia University Press, 1986. 270–95.

Miller, Naomi J. *Changing the Subject: Mary Wroth and Figurations of Gender in early modern England*. Lexington, KY: University of Kentucky Press, 1996.

Miller, Naomi J. and Gary Waller, editors. *Reading Mary Wroth: Representing Alternatives in Early Modern England*. Knoxville: University of Tennessee Press, 1991.

Milton, John. *Complete Poems and Major Prose.* Merritt Y. Hughes, editor. NY: Macmillan, 1985.

Minogue, Sally. "A Woman's Touch: Astrophil, Stella, and 'Queen Vertue's Court.'" *English Literary History* 63, 3 (1996): 555–70.

Montrose, Louis. "Celebration and Insinuation: Sir Philip Sidney and the Motives of Elizabethan Courtship." *Renaissance Drama.* New series 8 (1977): 3–36.

——. "The Elizabethan Subject and the Spenserian Text." *Literary Theory/Renaissance Texts.* Patricia Parker and David Quint, editors. Baltimore: Johns Hopkins University Press, 1985: 303–40.

——. "'The Place of a Brother' in *As You Like It*; Social Process and Comic Form." *Shakespeare Quarterly* 32 (1981): 28–54.

——. "'Shaping Fantasies': Figurations of Gender and Power in Elizabethan Culture." *Representations* 1, 2 (1983): 61–94.

Morenco, Franco. "Double Plot in Sidney's *Old Arcadia.*" *Modern Language Review* 64 (1969): 248–63.

Nashe, Thomas. *The Works of Thomas Nashe.* R.G. McKerrow, editor. Oxford: B. Blackwell, 1958.

Neale, J.E. *Elizabeth I and Her Parliaments 1559–1581* and *1584–1601* (2 volumes). NY: St Martin's Press, 1958.

Newcomb, Lori Humphrey. *Reading Popular Romance in Early Modern England.* NY: Columbia University Press, 2002.

——. "The Romance of Service: The Simple History of *Pandosto*'s Secret Readers. *Framing Elizabethan Fiction: Contemporary Approaches to Early-Modern Narrative Prose.* Constance C. Relihan, editor. Kent, OH: Kent State University Press, 1996. 117–39.

Olmsted, Wendy. "On the Margins of Otherness: Metamorphosis and Identity in Homer, Ovid, Sidney, and Milton." *New Literary History* 27 (1996): 167–84.

Orlin, Lena Cowen. "The Fictional Families of Elizabeth I." *Political Rhetoric, Power, and Renaissance Women.* Carole Levin and Patricia A. Sullivan, editors. Albany: State University of New York Press, 1995. 85–110.

Pask, Kevin. *The Emergence of the English Author: Scripting the Life of the Poet in Early Modern England.* NY: Cambridge University Press, 1996.

Patterson, Annabel. "No Meer Amatorious Novel?" *Politics, Poetics, and Hermeneutics in Milton's Prose.* David Loewenstein and James Grantham Turner, editors. NY: Cambridge University Press, 1990. 85–101.

Perry, Ruth. "Women in Families: The Great Disinheritance." *Women and Literature in Britain 1700–1800.* Vivien Jones, editor. Cambridge, UK: Cambridge University Press, 2000. 111–31.

Plowden, Alison. *The Elizabethan Secret Service.* NY: St. Martin's Press, 1991.

Poster, Mark. *Critical Theory of the Family.* NY: Seabury Press, 1978.

Prendergast, Maria Teresa Micaela. "Philoclea Parsed: Prose, Verse, and Femininity in Sidney's *Old Arcadia.*" *Framing Elizabethan Fiction: Contemporary Approaches to Early-Modern Narrative Prose.* Constance C. Relihan, editor. Kent, OH: Kent State University Press, 1996. 99–116.

Preston, Claire, "Sidney's Arcadian Poetics: A Medicine of Cherries and the Philosophy of Cavaliers." *English Renaissance Prose: History, Language, and*

Politics. Neil Rhodes, editor. Tempe, AZ: Medieval and Renaissance Texts and Studies, 1997. 91–108.

Quilligan, Maureen. "Completing the Conversation." *Shakespeare Studies* 25 (1997): 42–49.

——. "The Constant Subject: Instability and Authority in Wroth's *Urania* Poems." *Soliciting Interpretation: Literary Theory and Seventeenth-Century English Poetry.* Katharine Eisaman Maus and Elizabeth D. Harvey, editors. Chicago: University of Chicago Press, 1990. 307–35.

——. "Lady Mary Wroth: Female Authority and the Family Romance." *Unfolded Tales: Essays on Renaissance Romance.* George M. Logan and Gordon Teskey, editors. Ithaca: Cornell University Press, 1989. 257–80.

——. "Sidney and His Queen." *The Historical Renaissance: New Essays on Tudor and Stuart Literature and Culture.* Heather Dubrow and Richard Strier, editors. Chicago: University of Chicago Press, 1988. 171–96.

Quint, David. "The Boat of Romance and Renaissance Epic." *Romance: Generic Transformation from Chretien de Troyes to Cervantes.* Kevin Brownlee and Marina Scordilis Brownlee, editors. Hanover, NH: University Press of New England, 1985. 178–202.

Radway, Janice. *Reading the Romance: Women, Patriarchy, and Popular Literature.* Chapel Hill and London: University of North Carolina Press, 1984.

Relihan, Constance C. *Fashioning Authority: The Development of Elizabethan Novelistic Discourse.* Kent, OH: Kent State University Press, 1994.

Richetti, John J. *Popular Fiction Before Richardson: Narrative Patterns 1700–1739.* NY: Oxford University Press, 1969.

Roberts, Marthe. *Origins of the Novel.* Sacha Rabinovitch, translator. Bloomington, IN: Indiana University Press, 1980.

Robinson, Ian. *The Establishment of Modern English Prose in the Reformation and the Enlightenment.* NY: Cambridge University Press, 1998.

Ross, Deborah. *The Excellence of Falsehood: Romance, Realism, and Women's Contributions to the Novel.* Lexington: University Press of Kentucky, 1991.

Rowe, Kenneth. "Romantic Love and Parental Authority in Sidney's *Arcadia*." *University of Michigan Contributions in Modern Philology.* No. 4. Ann Arbor: University of Michigan Press, 1947.

Rudenstine, Neil L. *Sidney's Poetic Development.* Cambridge, MA: Harvard University Press, 1967.

Schlauch, Margaret. *Antecedents of the English Novel 1400–1600.* London: Oxford University Press, 1963.

Schwartz, Regina M. *The Curse of Cain: The Violent Legacy of Monotheism.* Chicago: University of Chicago Press, 1997.

Segre, Cesare. "What Bakhtin Left Unsaid: The Case of Medieval Romance." *Romance: Generic Transformation from Chretien de Troyes to Cervantes.* Kevin Brownlee and Marina Scordilis Brownlee, editors. Hanover, NH: University Press of New England, 1985. 23–46.

Shaver, Anne. "A New Woman of Romance." *Modern Language Studies* 21, 4 (1991): 63–77.

Shearman, John K.G. "Variety and Monotony." *Mannerism.* Harmondsworth: Penguin Books, 1967. 141–51.

Shuger, Debora Kuller. *The Renaissance Bible: Scholarship, Sacrifice, and Subjectivity*. Berkeley: University of California Press, 1994.

Smith, Hallett. "English Metrical Psalms in the Sixteenth Century and Their Literary Significance." *The Huntington Library Quarterly* 9 (1946): 249–71.

Spacks, Patricia Meyers. *Desire and Truth: Functions of Plot in Eighteenth-Century English Novels*. Chicago: University of Chicago Press, 1990.

Spencer, Theodore. "The Poetry of Sir Philip Sidney." *English Literary History* 12 (1945): 251–78. Rpt. in *Essential Articles for the Study of Sir Philip Sidney*. Arthur F. Kinney, editor. Hamden, CT: Archon Books, 1986. 31–59.

Spenser, Edmund. *The Faerie Queene*. A.C. Hamilton, editor. NY: Longman, 2001.

——. *The Yale Edition of the Shorter Poems of Edmund Spenser*. William A. Oram et al., editors. New Haven: Yale University Press, 1989.

Steinberg, Theodore. "The Sidneys and the Psalms." *Studies in Philology* 92, 1 (1995): 1–17.

——. "Spenser, Sidney, and the Myth of Astrophel." *Spenser Studies* 11 (1994): 187–201.

Stillman, Robert E. "The Politics of Sidney's Pastoral: Mystification and Mythology in *The Old Arcadia*." *English Literary History* 52 (1985): 795–814.

——. *Sidney's Poetic Justice: The Old Arcadia, Its Eclogues, and Renaissance Pastoral Traditions*. Lewisburg, PA: Bucknell University Press, 1986.

Stone, Lawrence. *The Crisis of the Aristocracy 1558–1641*. Abridged edition. London and NY: Oxford University Press, 1967, 1977.

——. *The Family, Sex, and Marriage in England 1500–1800*. London: Weidenfeld and Nicolson, 1977.

Tamen, Miguel. "Phenomenology of the Ghost: Revision in Literary History." *New Literary History* 29 (1998): 295–304.

Tennenhouse, Leonard. "Arcadian Rhetoric: Sidney and the Politics of Courtship." *Sir Philip Sidney's Achievements*. M.J.B. Allen et al., editors. NY: AMS Press, 1990. 201–13.

——. *Power on Display: The Politics of Shakespeare's Genres*. NY: Methuen, 1986.

Thirsk, Joan. "Younger Sons in the Seventeenth Century." *History* 54 (1969): 358–77.

Thomas, Keith. "Age and Authority in Early Modern England." *Proceedings of the British Academy* 62 (1976): 205–48.

Tillich, Paul. *The Dynamics of Faith*. NY: Harper & Row, 1957.

Trill, Suzanne. "Spectres and Sisters: Mary Sidney and the 'Perennial Puzzle' of Renaissance Women's Writing." *Renaissance Configurations: Voices/Bodies/Spaces, 1580–1690*. Gordon McMullan, editor. London: Macmillan, 1998. 202–22.

Tudor, Elizabeth. "Instructions for A.B." [Elizabeth's letter to Sir Francis Walsingham]. Cecil-Stamford-Towneley MS. The Pierpont Morgan Library. MA 664.

Waller, Gary. "The Countess of Pembroke and Gendered Reading." *The Renaissance Englishwoman in Print: Counterbalancing the Canon*. Anne M. Haselkorn and Betty S. Travitsky, editors. Amherst: University of Massachusetts Press, 1990. 327–45.

——. *Mary Sidney, Countess of Pembroke: A Critical Study of Her Writings and Literary Milieu.* Salzburg: Institut fur Anglistik und Amerikanistik, 1979.

——. "Mary Wroth and the Sidney Family Romance: Gender Construction in Early Modern England." *Reading Mary Wroth: Representing Alternatives in Early Modern England.* Naomi J. Miller and Gary Waller, editors. Knoxville: The University of Tennessee Press, 1991. 35–63.

——. "Mother/Son, Father/Daughter, Brother/Sister, Cousins: The Sidney Family Romance." *Modern Philology* 88 (1991): 401–14.

——. "The Rewriting of Petrarch: Sidney and the Languages of Sixteenth-Century Poetry." *Sir Philip Sidney and the Interpretation of Renaissance Culture: The Poet in His Time and in Ours.* Gary F. Waller and Michael D. Moore, editors. Totowa, NJ: Barnes and Noble Books, 1984. 69–83.

——. "The 'sad pilgrim': The Poetry of Sir Robert Sidney." *Dalhousie Review* 55 (1975–76): 689–705.

——. *The Sidney Family Romance: Mary Wroth, William Herbert, and the Early Modern Construction of Gender.* Detroit: Wayne State University Press, 1993.

——. "Struggling into Discourse: The Emergence of Renaissance Women's Writing." *Silent but for the Word: Tudor Women as Patrons, Translators, and Writers of Religious Works.* Margaret Patterson Hannay, editor. Kent, OH: Kent State University Press, 1985. 238–56.

——. " 'This Matching of Contraries': Calvinism and Courtly Philosophy in the Sidney Psalms." *Essential Articles for the Study of Sir Philip Sidney.* Arthur F. Kinney, editor. Hamden, CT: Archon Books, 1986. 411–23.

Warkentin, Germaine. "Robert Sidney's 'Darcke Offrings': The Making of a Late Tudor Manuscript." *Spenser Studies* 12 (1991): 46–50.

Warner, William B. "Formulating Fiction: Romancing the General Reader in Early Modern Britain." *Cultural Institutions of the Novel.* Deidre Lynch and William B. Warner, editors. Durham: Duke University Press, 1996. 279–305.

Wayne, Don E. *Penshurst: The Semiotics of Place and the Poetics of History.* Milwaukee: University of Wisconsin Press, 1984.

Weiner, Andrew D. "Spenser's *Muiopotmos* and the Fates of Butterflies and Men." *Journal of English and Germanic Philology* 84 (1985): 203–20.

Weiner, Seth. "The Quantitative Poems and the Psalm Translations: The Place of Sidney's Experimental Verse in the Legend." *Sir Philip Sidney: 1586 and the Creation of a Legend.* Jan Van Dorsten, Dominic Baker-Smith, and Arthur F. Kinney, editors. Leiden: Brill, 1986. 193–204.

Whigham, Frank. *Ambition and Privilege: The Social Tropes of Elizabethan Courtesy Theory.* Berkeley: University of California Press, 1984.

Williams, Raymond. *The Country and the City.* NY: Oxford University Press, 1973.

Winnicott, D.W. *The Child and the Family: First Relationships.* Janet Hardenberg, editor. London: Tavistock Publications, 1957.

——. *The Child and the Outside World: Studies in Developing Relationships.* Janet Hardenberg, editor. London: Tavistock Publications, 1962.

Wolfe, Linda. "The Strange Deaths of the Twin Gynecologists." *New York* September 8, 1975. 42–47.

Woolf, Virginia. *The Second Common Reader*. NY: Harcourt, Brace & Company, 1932.

Worden, Blair. *The Sound of Virtue: Philip Sidney's Arcadia and Elizabethan Politics*. New Haven: Yale University Press, 1996.

Wrigley, E. Anthony. "Reflections on the History of the Family." *Daedulus: Proceedings of the American Academy of Arts and Sciences* 106, 2 (1977): 71–79.

Wrightson, Keith. "Household and Kinship in Sixteenth-century England." *History Workshop* 12 (1981): 151–58.

Wynne-Davies, Marion. "'For *Worth*, Not Weakness, Makes in Use but One': Literary Dialogues in an English Renaissance Family." *"This Double Voice": Gendered Writing in Early Modern England*. Danielle Clarke and Elizabeth Clarke, editors. NY: St. Martin's Press, 2000. 164–84.

———. "'So Much Worth': Autobiographical Narratives in the Work of Lady Mary Wroth." *Betraying Ourselves: Forms of Self-Representation in Early Modern English Texts*. Henk Dragstra, Sheila Ottway, Helen Wilcox, editors. NY: St. Martin's Press, 2000. 77–93.

INDEX

Printed in the United States
By Bookmasters